Vehicle Fabrication in G.R.P.

Peter Child
MIMI

BSP PROFESSIONAL BOOKS
OXFORD LONDON EDINBURGH
BOSTON PALO ALTO MELBOURNE

Copyright © Peter Child 1987

All rights reserved. No part of this
publication may be reproduced, stored
in a retrieval system, or transmitted,
in any form or by any means,
electronic, mechanical, photocopying,
recording or otherwise without
the prior permission of the
copyright owner.

First published 1987

British Library
Cataloguing in Publication Data
Child, Peter
  Vehicle fabrications in G.R.P.
  1. Plastics in automobiles   2. Polyesters
  3. Glass reinforced plastics
  I. Title
  629.2'32   TL154

ISBN 0–632–01871–2

BSP Professional Books
Editorial offices:
Osney Mead, Oxford OX2 0EL
  (*Orders:* Tel. 0865 240201)
8 John Street, London WC1N 2ES
23 Ainslie Place, Edinburgh EH3 6AJ
52 Beacon Street, Boston
  Massachusetts 02108, USA
667 Lytton Avenue, Palo Alto
  California 94301, USA
107 Barry Street, Carlton
  Victoria 3053, Australia

Set by V & M Graphics Ltd.,
Aylesbury, Bucks
Printed and bound in Great Britain by
Billing & Sons Ltd., Worcester

Contents

Preface	iv
Acknowledgements	iv
Introduction	v
1. Glass Reinforced Polyester	1
2. Design	7
3. Mould Making	21
4. Taking Moulds from Vehicles	28
5. Preparing Moulds	35
6. Laying up	38
7. Reinforcement Process	49
8. Release from the Mould	53
9. Fixing Bodyshells	55
10. Trimming and Faults	60
11. Painting a GRP Bodyshell	65
12. Material Developments	70
13. Future Trends	79
14. Repair of a GRP Bodyshell	83
15. Safety	86
Glossary	89
Index	90

Preface

The advance of technology has been impressive, if not totally overwhelming, since the Second World War. The full application of this knowledge into the construction of everything that is needed on this planet has been rapid due to the financial pressures to use new developments to make life safer and more comfortable in the most effective financial way.

The use of 'plastics' in every form has really reached out for the motor industry. The application of techniques learned in the aerospace industry has fitted neatly into the demands of the automotive field. Lightweight, immensely strong, almost non-destructible materials have been well received by the automobile engineer who has been able to flex his ideas and develop still further.

The fact that top quality manufacturers are using glass reinforced polyester because of its strength and lightweight characteristics shows how close aviation and high performance vehicles have become. As more and more mass-produced cars use 'plastics', so road vehicles will be safer and easier to maintain, and have a useful lifespan that can only be hoped for at present.

Peter Child

Acknowledgements

I gratefully acknowledge all the help and kind consideration shown to me by the following, without whose help and co-operation this book could not have been written.

Aston Martin Lagonda Ltd
Copperleaf Cars Ltd
DuPont International
ICI
Scott Bader Ltd
A. M. Stevens Esq.
J. Masters Esq.

Introduction

Within the last ten years enormous strides have been made in glass reinforced polyester technology. So much so, that the growth of its use has spread throughout the motor industry to Aston Martin Lagonda and Rolls-Royce. The strength, durability, and weight-saving characteristics have made this material invaluable to the motor industry. For instance, Ford Motor Company, Transmission and Chassis Division in the USA, have developed a carbon glass fibre composite drive shaft for the 1985 Econoline van. Claimed to be the world's first high volume composite drive shaft, it is a filament wound in a five-minute cycle by Hercules Aerospace who use a 20% carbon to 40% glass reinforced vinyl ester resin. This component replaces the previous two-piece steel shaft unit and has reduced the weight from 31 lb to 19 lb – a saving of 12 lb.

This technology, developed for the aircraft industry, is now widely available for the vehicle and marine industries. The ever-growing interest in specialist cars has helped promote the situation further. Lotus, with their excellent bodyshells, have helped to develop the GRP industry for vehicles and they have been at the forefront with development and testing. With TVR and Reliant making more and more vehicles to meet the demand, the added impetus has helped the resin manufacturers to research further, and now the GRP motor vehicle is both here to stay and is certain of lasting much longer than its steel counterpart. The days of the all-steel motor vehicle are numbered.

An unsaturated polyester resin is produced by a chemical reaction between organic acids and glycols. The acid component consists of saturated and unsaturated acids. A monomer is added as solvent. The monomer has a twofold task:

1. To dissolve the polyester resin.
2. To react during curing with the double links in the unsaturated acid, and thus crosslink the polyester molecules. The monomer is therefore not a solvent in the ordinary sense, but a vital component in the curing process.

The most usual raw materials are:

- Unsaturated acid: malic anhydride.
- Saturated acid: phthalic anhydride.
- Glycol: propylene glycol.
- Monomer: styrene.

Glass fibre reinforced unsaturated polyester: GRP

As a group of materials GRP belongs to the composites since two or more separate components are involved, these being a liquid thermosetting resin and a solid fibre-like material. The matrix in GRP is unsaturated polyester resin, while the reinforced material consists of glass fibre in the shape of continuous or chopped roving, chopped strand mats, woven roving or cloth. The unsaturated resin is transformed into a solid state by adding a catalyst, usually organic peroxide. In order to speed up the process, the catalyst must be activated either by applying heat or by an accelerator. The catalyst triggers off a chemical reaction known as copolymerisation: the monomer reacts with the unsaturated groups in the polyester chains forming a crosslinked polymer, a so-called thermoset. When combined with a reinforcement material, this is known as a reinforced thermoset.

The beginnings of GRP

GRP is a young material. The first products were made in the United States in the early 1940s for maritime purposes. Jotun manufactured its first polyester resin in 1954. There has since been a dynamic development, and today Norway has the world's highest GRP consumption per capita. Jotun has played an active part in this trend, at home and on an international scale. These years have seen the successive development of various polyester resin qualities adapted to a variety of production methods and a multitude of product purposes. Inventive and creative people have realised that GRP is literally the material of boundless potential, and are now utilising it to an ever-increasing extent.

GRP and the sea

Originally GRP was a construction material with a maritime slant. Broadly speaking, other application fields now predominate. Nevertheless, the maritime application was developed first for the pleasure craft industry and then, as the store of experience grew and technology improved, GRP was employed in larger maritime constructions.

The most interesting features of GRP are its low weight in relation to mechanical strength, and its easy maintenance; it neither corrodes, nor weakens at low temperatures. Today GRP is widely used in lifeboats, passenger ships, fishing vessels and naval craft. In recent years a new technique has been evolved for larger vessels – the sandwich method: a lightweight core material such as PVC or urethane foam is overlaminated with GRP.

The design concept of any vehicle has to meet the parameters of use and ease of build. In a steel car with pressing tools the ease of build is extremely important. However, in GRP, there is no restriction. The design of any vehicle is not hampered by cost nor by structure limitations. Any style, shape or configuration of cosmetic panels can be achieved by the use of a pattern and then a mould. GRP is so versatile that any shape, no matter how intricate, can be constructed.

1

Glass Reinforced Polyester

The material that is often called 'fibreglass' is in fact 'glass fibre reinforced plastic', or 'glass fibre reinforced polyester', or GRP. It is a lightweight, immensely strong and durable material. It can be fabricated into any shape and can be self-coloured or opaque. In the marine industry boat hulls up to 60 metres long have been constructed in one piece. The user or manufacturer of an item to be made in GRP actually makes the plastic as the item is moulded. GRP is a composite of resilient and durable resin with a very strong fibrous glass. The resin is the main component and is usually a polyester resin. It is supplied in a viscous form and is activated by a catalyst into a rigid, solid form.

Polyester resins are reinforced with glass fibres and using a mould, a manufacturer lays down resin mixed with catalyst and glass mat in layers. By rollering this composite, complete saturation of the mat by the resin bonds the whole lay-up into a solid form material.

Plastics

All plastics are man-made materials which can be moulded into shapes. The first plastic was developed in England by Alexander Parkes in 1862. This material was called Parkesine and was the forerunner of celluloid. Since then a variety of plastics have been developed commercially, most of them over the last quarter century. They consist of a wide range, from phenol formaldehyde (PF), which is a hard thermoset material, to PVC (polyvinyl chloride) which is a soft, tough thermoplastic material. Plastics also exist in various forms. They can be bulky, solid materials, rigid or flexible foams. They also exist in sheet or thin film form.

Plastics all have one common property. They are all composed of macro-molecules which are large, chain-like molecules consisting of many repeating units. These molecular chains are called polymers. Polymers appear in diverse materials such as paint, man-made fibres and nylon, to name but a few. Many polymers occur naturally, e.g. cellulose (wood and cotton, used in the manufacture of nitrocellulose paints for finishing vehicles) and rubber. Man-made polymers are called synthetic resins. These are made from oil or coal tar.

Reinforced plastics

Resins are usually combined with fillers to improve the physical properties of the plastic and to reduce the cost in manufacture. These fillers can be any powdered mineral, china clay or wood flour. High strength plastics can be made by reinforcing the resin with layer after layer of paper or fabric. The result is called a laminate. Most reinforced plastics consist of glass fibre reinforced polyester resin.

Table 1 shows the comparative properties of glass reinforced plastics.

Vehicle Fabrications in GRP

Table 1

Material	Glass content		Specific gravity	Tensile strength MPa	Tensile modulus GPa	Specific strength MPa
	% vol.	% weight				
Polyester/rovings	54	70	1.9	800	30	400
Polyester/cloth	38	55	1.7	300	15	200
Polyester/mat	18	30	1.4	100	7	70
Mild steel	–	–	7.8	310	200	40
Duralumin	–	–	2.8	450	70	150
Douglas fir	–	–	0.5	75	13	150

Glass fibre

Glass fibre is made by rapidly drawing and cooling molten glass. Although the process has been known for thousands of years, glass fibres of sufficient fineness and consistency were not developed commercially until fifty years ago.

There are two types of glass fibre:

1. A coarse staple glass fibre that is widely used as an insulator. This material is not suitable for use as a plastics reinforcement.
2. A type that consists of continuous filaments which, after drawing, are bundled up together to form strands. These are made into rovings or are weaved into glass cloth.

Glass fibre is a very strong material: the ultimate tensile strength of a freshly drawn single glass filament of 9 to 15 microns (25 microns to one thousandth of an inch approximately) is about 3.5 GPa. It is made from raw materials that are readily available, it will not support combustion and is resistant to chemical attack. Glass fibre is the ideal reinforcing material for plastics. The material as we know it today became commercially available during the Second World War, and is described as a low pressure or contact resin. These materials heralded the introduction of the GRP that has now been developed and is currently on offer.

Carbon fibre

Carbon fibre is a new development in the field of reinforcement and was developed for the field of aviation by the Royal Aircraft Establishment at Farnborough in the '60s. The technicians there developed a method for manufacturing a highly crystalline fibre. Normally, carbon fibre has a strength and modulus greatly superior to glass fibre. This material is costly but the application of this technology into aircraft, both military and commercial, will without doubt encourage its use in the motor industry. Certainly the racing car designers are using more and more in an attempt to reduce weight and give greater strength to the structure.

Glass Reinforced Polyester

Polyester resins

Crystic resins are unsaturated polyester resins. This means that they are polyester resins capable of being cured from a liquid to a solid state when subjected to the right conditions. They differ therefore from a saturated polyester such as Terylene which cannot be cured in this way. However, it is usual to refer to unsaturated polyester resins as polyester resins, or simply polyesters.

Most crystic resins are liquids and they consist of a solution of a polyester in a monomer, which is usually styrene. The styrene performs the vital function of enabling the resin to cure from a liquid to a solid by crosslinking the molecular chains of the polyester, without the evolution of any by-products. They can therefore be moulded without the use of pressure; these are known as contact or low pressure resins.

The molecular chains of the polyester can be represented as follows:

$$- A - B - A - B - A - B -$$

With the addition of — S —, and in the presence of a catalyst and accelerator, the styrene crosslinks the polymer chains to form a highly complex three-dimensional network as follows:

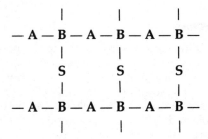

The polyester resin is then said to be cured. It is now a chemically resistant hard solid. The crosslinking or curing process is called polymerisation. It is a non-reversible chemical reaction.

Applications

Crystic polyester resins have many uses, but generally the application of these resins is in the use of glass fibre reinforced polyester laminates and mouldings.

Table 2 shows the typical physical properties of GRP with different types of glass fibre reinforcement.

Gelation and hardening

Storage

Storage life of liquid polyesters is uncertain due to their instability. After

Vehicle Fabrications in GRP

Table 2

Properties	Chopped mat	Woven rovings	Satin cloth	Cont. rov.
Glass content (% weight)	30	45	55	70
(% volume)	18	29	38	54
Specific gravity	1.4	1.6	1.7	1.9
Tensile strength (MPa)	100	250	300	800
Tensile modulus (GPa)	8	15	15	40
Compressive strength (MPa)	150	150	250	350
Bend strength (MPa)	150	250	400	1000
Modulus in bend (GPa)	7	15	15	40
Impact strength Izod (Kj/m^2)	75	125	150	250

several months or even years they will set into a rubbery gel. This will occur even at room temperatures. This period is known as the storage life or shelf life of the resin and it varies from one type of polyester to another.

The shelf life is considerably reduced at temperatures greater than 25°C, or if the resin is stored in glass containers and exposed to ultra violet light. Most crystic resins have a storage life in the dark at 20°C of at least six months, and in some cases more than a year. Pigmented crystic resins have a shelf life of three months.

Catalysts and accelerators

In order to produce a moulding or GRP laminate the polyester resin must be cured. This is the name given to the overall process of gelation and hardening. This is achieved by the use of a catalyst and the application of heat, or at normal room temperature by using a catalyst and an accelerator.

Catalysts for polyester resins are normally organic peroxides. Pure catalysts are chemically unstable and are liable to decompose. They are always supplied in a liquid or paste form dispersed in a plasticiser.

It is essential to choose the correct type of catalyst and accelerator as well as to use the correct amount, if the optimum properties of the final cured resin or laminate are to be obtained. If the correct and recommended catalyst is used in the mixing and forming of GRP laminations then the matrix will achieve its maximum strength, durability, chemical resistance and stability.

Catalysts and accelerators must never be mixed together as a violent reaction occurs that can be explosive.

Pre-accelerated resins

Many crystic resins are supplied with a built-in accelerator system, controlled to give the most suitable gelling and hardening characteristics for the fabricator. Pre-accelerated resins need only the addition of a catalyst to start the curing process at room temperature.

Glass Reinforced Polyester

The curing reaction

The cure of a polyester resin will commence as soon as a suitable catalyst is added. The speed of the reaction will depend on the resin and the activity of the catalyst. Without the addition of the accelerator, heat or ultra violet radiation, the catalysed resin will have a pot life of hours or even days. This rate of cure is too slow for practical purposes so that in room temperature conditions it is usual to add an accelerator to speed up the reaction. The quantity of accelerator added will control the time to gelation and the rate of hardening.

A limited pot life of a catalysed resin may be undesirable and in these conditions it is advisable to add the required quantity of accelerator to the resin first of all. This will then remain usable for a period of time extending into weeks. Small quantities of this material can then be catalysed as and when required.

The reaction is exothermic and there is a temperature rise. There are three phases in the curing reaction:

1. The gel time. This is the period from the addition of the accelerator to the setting of the resin to a soft gel.
2. Hardening time. This is the time from the setting of the resin to the point when the resin is hard enough to allow the moulding or laminate to be withdrawn from the mould.
3. Maturing time. This may be hours, days or even weeks depending on the resin and curing system, and is the time taken for the moulding or laminate to acquire its full hardness, chemical resistance and stability. Maturing will take place at room temperature or it can be accelerated by post-curing.

Factors affecting gel time

The following factors influence the gel time of crystic resin and therefore the final state of cure:

* Catalyst content. The less catalyst used, the longer the gel time. Insufficient catalyst leads to undercured mouldings.
* Accelerator content. The less accelerator used the longer the gel time. When there is insufficient accelerator to activate the catalyst the resin may remain undercured, or harden too slowly.
* Ambient temperature. The lower the temperature the longer the gel time. Curing below 15°C can lead to undercure.
* Bulk of resin. The larger the bulk of the resin, the shorter the gel time. For example a 25 mm cube casting will set faster than a 2 mm laminate using the same formulation.
* Loss of monomer by evaporation. It is essential to have enough monomer in the resin for adequate polymerisation. When laminating large areas therefore it is recommended that the resin be made to gel quickly.
* Choice of fillers when these are used. Most fillers lengthen the gel time.
* Pigment content. Certain pigments lengthen the gel time, others shorten it. The effect of pigments which are not specifically recommended for polyester resins should therefore be determined before use.

- The time-lag between the addition of the catalyst and the addition of the accelerator. The longer the catalyst resin has been stored the shorter the gel time.
- Presence of inhibitors. These are compounds, small traces of which are sufficient to poison the polymerisation reaction, and may prevent full cure altogether. The most common inhibitors are phenols, phenol formaldehyde resin dust, sulphur, rubber, copper and copper salts, most forms of carbon black and methanol.

Pigments

No more pigment than is absolutely necessary to achieve the desired depth of colour or opacity should be added. Up to 10% by weight of a suitable polyester pigment paste can be used. Many resins are available already pigmented.

Fillers

Mineral fillers gained a poor reputation when they were first introduced to the reinforced plastics industry, mainly due to the crudely ground limestone materials. These 'fillers' were used in excessive loadings solely to reduce the cost of mouldings, without attention being paid to the serious deterioration in mechanical strength and chemical resistance properties which they caused. Now, however, several mineral fillers are available which are particularly suitable in GRP mouldings. In fact, it is now recognised that the cost-saving resulting from the use of a filler is of secondary importance compared with the improvement of properties which can be achieved.

Surface treated calcium carbonate fillers, particularly crystalline types, are now widely used in the reinforced plastics industry and the leading manufacturers can recommend grades for many applications. As a general rule filler content should be kept as low as possible. If finely divided filler has to be used, not more than 25% of resin weight of mineral filler should be included.

Mixing

All component materials must be thoroughly dispersed in the resin, since inadequate mixing can only lead to faulty moulding. The order of mixing the catalyst and accelerator will depend on the particular application and the curing system used.

The future

Future developments in all aspects of GRP are as assured as they are exciting. The motor industry world-wide will develop and use this remarkable material to the ultimate.

Plastics have so many uses and the level of technical control over the material makes it a necessary and basic tool for development now and in the next century. Controllability, reliability and above all total consistency to a known performance, make the use of these materials a certainty for an indefinite period of time to come.

2
Design

For the construction of a GRP vehicle experience has proved that it is necessary to combine both GRP and steel. The three leading exponents of GRP vehicle construction, i.e. Lotus, Reliant and TVR, all use a steel chassis as a platform or backbone to a GRP body structure.

Although these three manufacturers all build sports cars, the same principles apply to a cross-country vehicle or a commercial van or pickup. The steel chassis is used to support the mechanical parts and the GRP bodyshell will carry everything else.

The undoubted strength of GRP enables the material when correctly formed to carry doors and tailgates with ease. These units on a vehicle are subject to heavy loads and stresses, and support for these must be correctly set in the GRP lay-up. Torsional strength may be built in by the use of box sections in GRP. Areas such as door sills and transverse ends of the floor pan will give added stability to the structure.

Proper fabrication and fixing of the bulkhead will reduce any 'scuttle shake' to a minimum. With some steel support for steering and brake mechanism, any movement on the bulkhead will be eliminated. By designing a deep transmission tunnel throughout the length of the floor pan a great contribution will be made to the overall strength of the structure.

It is important that careful thought is given so that the ultimate strength of the bodyshell is designed in at the very outset. By providing deep sills, transmission tunnel and transverse end box-sections within the design, a robust vehicle, capable of giving very long-term service, will result.

When creating a design it is vital to think in terms of GRP and what it can accomplish, rather than thinking of the structure in another material such as steel or wood, and wondering how to construct that item in GRP. For example, all edges should have a generous radius – GRP does not perform well over sharp corners; as GRP is prepared in a mould, at least a 2 degree taper is necessary to draw the fabrication from the mould. If this cannot be accomplished then a two or more piece mould must be constructed.

For low production levels of any motor vehicle, the hand lay-up method is usually the one chosen for its ease and simplicity. Application of materials by the spray method is very effective but expensive to put into operation with regard to the number of units planned for production.

The use of GRP gives the designer the design freedom that virtually no other construction material can offer. For example, double curvature profiles on panels: these are very difficult to work in conventional materials such as steel or aluminium. Most complex lines and curves may be transmitted directly from the sculptured 'plug' or male mould directly through the female mould to the final GRP bodyshell. Exotic designs and styles dreamed up by the stylists can be put into immediate construction. A visit to any of the major European Motor Shows will prove the case. GRP is able to cope with the stylists' wildest inventions.

No matter how simple or how complex the shape, when using GRP certain characteristics need to be stated:

1. When laying up avoid poor fibre distribution and the inclusion of air bubbles.
2. All radii should be generous and preferably 6 mm minimum.
3. Undercuts can be moulded provided that the mould is split.
4. For a one-piece mould a minimum of 2 degrees of taper is necessary to draw the laminate out of the mould.

The thickness of the GRP section is normally a minimum of about 0.75 mm on hand lay-up. There is no upper limit on thickness, but as a guide, a single gel coat followed by three lay-ups of $1\frac{1}{2}$ oz chopped mat will give a total thickness when fully cured of about 4 mm. This is ample for the cosmetic panels on a GRP bodyshell. When producing wet on wet lay-ups, care must be taken that the exothermic heat that is developed does not cause damage to the mould surface. The three laminations giving 4 mm thickness will not cause any undue exothermic reaction if the operation is carried out at normal working temperature – an operating window of between 65°F and 72°F. It is possible when working at these levels to bond in stiffeners or ribs, and provided not too much resin is used and the mat is correctly wetted out, then again, exothermic reaction will be kept to a minimum.

In the design of the moulding it is important to recognise the safety factors that are necessary if the vehicle is to function correctly and to maintain the level of safety indefinitely. Safety factors can be based on an assessment of the predicted load that the bodyshell will suffer in normal use. This normal use must be stressed, as cross-country vehicles or commercial load-carrying vehicles will bear a much heavier loading than a normal road car used in a responsible manner.

Table 3 gives a guide to the designer, bearing in mind that the figures represent, even at minimum, a considerable plus factor on what might otherwise be desired if working in steel or other materials. This is because so much of the strength of the laminate depends on the operator. One could consider this table to represent a worst case situation. The maximum figures given are for a high safety margin for vehicles used in normal road-going conditions and impact loads must be on an ascending scale from either end of the three-cell construction.

Table 3

Static or short-term loads	=	Safety factor minimum 2, maximum 3
Static long-term loads	=	Safety factor minimum 4, maximum 5
Variable loads	=	Safety factor minimum 5, maximum 6
Repetitive loads	=	Safety factor minimum 6, maximum 7
Load reversal	=	Safety factor minimum 6, maximum 7
Impact loads	=	Safety factor minimum 12, maximum 14

Design

Fig. 1 A box-section chassis manufactured in mild steel with Ford Cortina mechanical parts fitted. This is the chassis designed by Copperleaf Cars Ltd for the Monaco sports car.

The larger the overall design of vehicle the greater the need for strengthening and reinforcement. The flexing of a steel chassis and bodyshell can be quite severe and although it is desirable to allow a minimum of flex it must be well and correctly supported. This unequal flexing must be accounted for in the fixing of the GRP body shell to the steel chassis. If this is not carried out properly then stress raising areas will abound. Some form of flexible mounting and insulation between GRP and steel is necessary. If this is not satisfactorily in place then not only will stress at specialised points be present, but a great deal of road noise and shock will be directed into the passenger compartment.

All vehicles should have a mild steel box-section chassis carrying the suspension in running gear. All the mechanical parts of the vehicle should be fixed directly to the chassis, so therefore only the body structure is manufactured in GRP. For the safety and collision aspects it is better to combine a rigid chassis structure with a monocoque GRP body shell. If an insulator of rubber or similar is used to separate the monocoque body shell from the chassis at the support and resting points the vibration and road shock can be greatly reduced.

A simple ladder chassis constructed with box-section mild steel is the ideal for 'one-off' or prototype construction (Fig. 1). This, coupled with a strong and rigid shell of GRP, combines to make a motor vehicle of immense strength but still lightweight for performance and economy (Fig. 2).

In the design of any motor vehicle there has to be an inner structure of floor pan, bulkheads, wheel arches and the front and rear under panels, with the cosmetic main frame of roof, front and rear wings, with front and rear cross panels. Boot, bonnet and doors are obviously separate.

The ideal way to construct a monocoque in GRP is to make two major moulds, one for the floor pan and bulkheads and the other for the cosmetic

Vehicle Fabrications in GRP

Fig. 2 The Monaco designed by Copperleaf Cars Ltd and made in one piece using 1½ oz mat in three laminations.

panels. Then with the use of fixing jigs, join the two together during the 'green period' of cure, using mat in strips.

The design of a vehicle is then divided into two basic major components, and a scale drawing showing these clearly must be drawn up.

Having completed the design work, a full-size mock up of the floor pan and body shell must be constructed. For the floor pan and bulkhead a very satisfactory structure can be made using 3 inch × 2 inch wood beams supporting sheet contiboard, which is a plastic-coated wood laminate. This can be cut and screwed into the beam frame to the dimensions of the drawing and will secure to take a female mould with very little work. The bodyshell plug must be constructed (Fig. 3) on a wooden base using formers over bulkheads to give the shape of the vehicle (Fig. 4). This must be covered in hardboard (Fig. 5) to hold the clay used for shaping. This is then applied to the full plug and shaped into the line and style of the vehicle. Templates must be constructed to

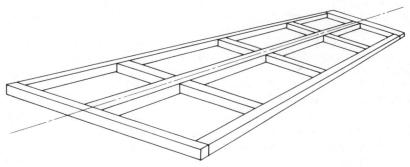

Fig. 3 A wooden main frame with a drawn centre line.

Design

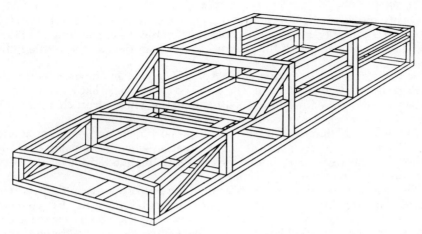

Fig. 4 A wooden outer frame assembled on the wood base.

take shapes and contours to offer up to either side for accuracy in the finished shape.

This phase of the operation is long and time-consuming. Once accomplished, however, the mould can be taken from which hundreds of vehicle bodies can be produced.

To take a mould from clay, great care must be excercised to ensure that there are no re-entry curves or shapes that will cause problems when the mould is released. If these are present then they must be taken separately from the mould and used in conjunction with the main mould as a secondary mould. This has to be bolted up before a shell is moulded from the female, and broken down before the male can be released.

The relationship of the clay to the design can be a matter of interpretation of the design form, and alterations of an aesthetic nature can be carried out at this stage.

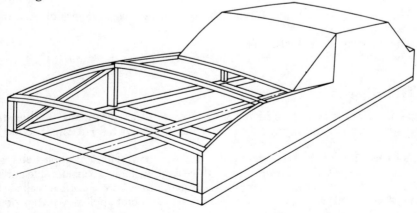

Fig. 5 Wooden outer frame partially covered in hardboard or plywood.

11

Vehicle Fabrications in GRP

Once a mould is cast then alterations are virtually impossible. When a clay mock-up is complete, imagination plays a part in the eye of the designer. How will it look painted? – with the glass in? – sitting on four wheels at the right height?

So many areas are deceptive. As an example, look at any modern motor car without its rear bumper in place. The back end always looks like a vast expanse of panel work. Once the bumper is fitted then the whole aspect changes – the car will appear balanced and the design will be harmonious.

Any vehicle design in solid will look disproportionate and not as light and sleek as one would wish, but the transformation when glass, paint, chrome, lights and wheels are added is total. Proceed with this in mind at every stage of the construction of the plug.

The feasibility of design revolves around the simple approach. Complicated shapes, although they can be achieved with the flexibility of GRP, are often not as stylish and elegant as they might be. Simple shapes are both easy to look at and easy to construct. Vehicle shapes should be functional as well as aesthetic. Nothing looks worse than a vehicle shape covered in 'add-on' items. Take the American designs in the late 'fifties, where wing fins, large areas of chrome, and lights of all varieties on pods made the vehicles look like large moving Christmas trees. When the designers moved on to clean and uncluttered shapes, the vehicles became much more attractive. The Ford Mustang Mark 1 is a good example of a clean-cut design which is extremely good-looking because of its simplicity.

The strength of GRP is considerable and the comparison with mild steel of comparable thickness is most impressive. The lightness of the cured structure is a great advantage, as Table 1 clearly shows. Impact strength, supportive strength and resistance to both vibration and fire (up to Class 1 with fire retardant resin), are impressive and the resistance to ultra violet light and water, in any form, ensures that GRP will outlast any other form of material used in the construction of motor vehicles.

To outline the reinforcements available for use in GRP composition one can look into three areas:

1. The use of polyurethane foam as a sandwich between layers of mat and · resin (Fig. 6).
2. The bonding in of steel support.
3. The addition of rovings or extra mat to strengthen the laminate.

Polyurethane foam

Rigid urethane foam is regarded as the most efficient means of thermal insulation available for a given thickness, and in sandwich form construction it gives high strength at relatively low weight.

Its formability as an expanding polyurethane foam will fill complex shapes and cavities completely. Its adhesion at the foaming stage is considerable and it will adhere firmly to most materials. It can be poured into a double wall such as a bulkhead or it can be cut to shape and resin and mat applied to either side.

Design

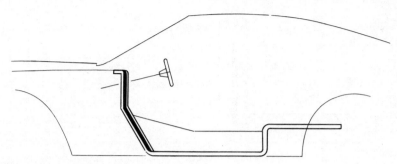

Fig. 6 A side view of a floor pan and bulkhead with foam laminated into the bulkhead.

Steel support

A steel support can be used in any section and depends on the strength required. A $3/16$ inch mild steel bar can be incorporated and this will provide rigidity (see Fig. 7). However, an angle section fully bonded in will resist all bending movement. A full box-section, such as 1 inch × 1 inch, will provide total rigidity. All steel inserts must be completely clean and free from grease. Heavy scoring on the surface will help the resin to adhere. In the case of the angle section, holes can be drilled through prior to laminating, which makes the bond even stronger as resin migrates through the holes from one side to the other (see Fig. 8).

Rovings and extra laminations

These sheet and strip materials can be added during or subsequent to the lay-up. Too much application of resin at this time can cause severe exothermic

Fig. 7 A-post windscreen pillars reinforced with steel bar.

Vehicle Fabrications in GRP

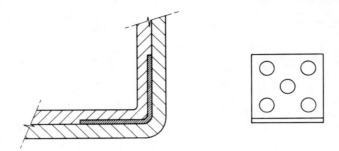

Fig. 8 Right-angled steel plate drilled out to allow resin migration and bonded into GRP fabrication.

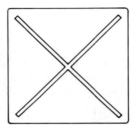

Fig. 9 A bonnet with cross reinforcement.

Fig. 10 A bonnet with added strength through the construction of returns.

reaction and the piece can become very hot, and in the extreme, can ignite. Proceed with caution here and ensure that the minimum of catalyst is used in the mix and that the minimum of resin is laid in to saturate the mat. Allow mat to 'wet up' and roller gently to achieve best results.

To use reinforcement as a principle of design is the right way to proceed. To conceive a shape in GRP and then, when it is constructed, to think of ways of reinforcing it, is a retrograde step. Design the reinforcement within the overall conception. For example, a bonnet panel with three laminations of $1\frac{1}{2}$ oz mat will be strong but too flexible. Add a 1 inch return all the way round the edge and the rigidity is increased dramatically. Add a simple cross of mild steel bar, welded together at the cross, bonded in after the third lay-up with strips of

Design

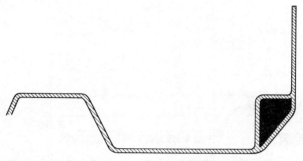

Fig. 11 Cross-section through floor pan and transmission tunnel showing boxed in door sill filled with foam.

mat, and the bonnet will be totally rigid (see Figs. 9 and 10). Equally, the box-sections in doors or sills that have been filled with polyurethane foam will dramatically strengthen the structure (Fig. 11).

Understand where the reinforcement is to be placed and design accordingly, using the materials and the technology that are available. Areas of loading or stress, such as bulkheads, door sills, A and B posts and rear supports, or C posts, need careful attention to give the strength necessary for a rigid monocoque vehicle.

In this design concept a point to keep in mind is the need for combining the maximum strength necessary to achieve rigidity with the lightest film weight possible. In other words, do not go on laying up more and more mat and resin on the door skins when the strength should lie in the door *frames*. Three sheets of $1\frac{1}{2}$ oz mat is sufficient for any door skin on a correctly designed door frame. Support the door frame with steel inserts and polyurethane foam and bond the skin to the box-section, as in Fig. 12.

In the event of an accident, the implosion factors have to be considered. All modern motor vehicles have to pass the DTP Type Approval tests. The UK

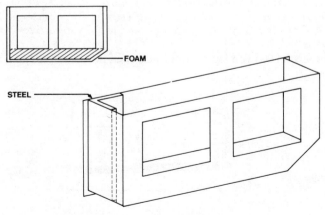

Fig. 12 Door construction.

15

Vehicle Fabrications in GRP

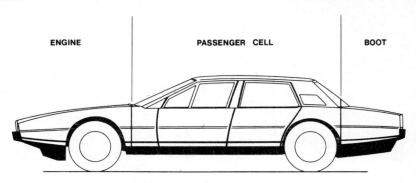

Fig. 13 Three cell construction.

Approval Test is not as comprehensive as that of the EEC, but nevertheless all motor manufacturers have to submit their vehicles to the most stringent tests. Part of this testing consists of impact damage up to speeds of 50 mph. Also, side and rear impact tests are carried out and the sustained damage is carefully analysed.

Passenger cell safety is the main point of interest and it is always expected that a vehicle will implode and absorb the impact. A totally rigid vehicle would not absorb the energy of a crash, much to the detriment of the occupants. Therefore all vehicles are expected to collapse on impact to a certain level, but allow the passengers a sizeable safety margin.

In what is known as a three compartment configuration, i.e. engine compartment, passenger cell, rear compartment, the front and rear should be designed to collapse. Side implosions will expose the need for stiff A and B posts and strong door support after initial cosmetic collapse (see Figs. 13 and 14).

The collapse on impact must be progressive with, in the case of front impact, the damage ceasing at the point of the passenger bulkhead, and with steering and dash assemblies collapsing down to avoid chest and leg injuries to the driver and passenger. Similarly for a prototype or 'one-off', these factors must be borne in mind and designed into the vehicle for the safety of all concerned. Weaker areas must be present at the front and rear of the vehicle to allow initial collapse, with the structure being stronger and more supportive nearer to the front and rear bulkheads.

Programmed collision collapse can be designed into the structure by using a

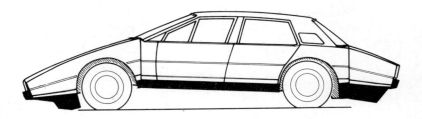

Fig. 14 The engine compartment and the boot collapse down on impact.

Design

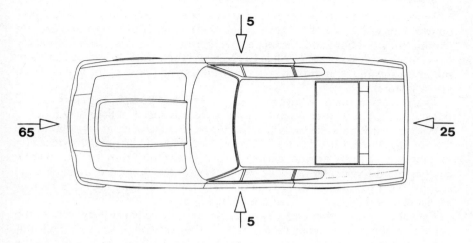

Fig. 15 UK damage distribution expressed in percentage.

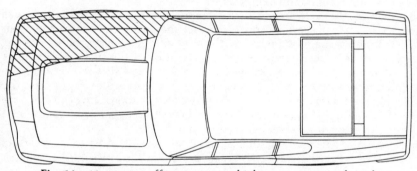

Fig. 16 40 per cent offset equates to highest proportion of accidents.

forward light box-section that will collapse up to the main chassis structure carrying the engine and running gear. This also can be employed at the rear up to the rear axle location and suspension points.

GRP laminations can consist of three layers of mat on the front and rear panels rising to four or five at the bulkhead where further layers of mat are used to bond the floor pan and bulkhead to the cosmetic panels, and this could increase that number to a total of eight to ten including both sides of the bulkhead fixing.

This area of the vehicle is very substantial, and certainly the statistics show that over 60% of vehicle damage is to the front; this is therefore the way to proceed to achieve the necessary design strength.

The average accident

Accidents can be categorised statistically, and for the UK the distribution is as shown in Fig. 15. For Europe it is much the same, with only slight percentage points in variation.

Vehicle Fabrications in GRP

By far and away the most frequent damage mode is the so-called three-quarter frontal (Fig. 16). With correct and well thought out design, and bearing in mind that the above statistics are a true representation of what is most likely to occur in a collision, it is possible to construct the vehicle in such a way as to minimise any injury to the driver and passengers, and protect the vehicle from serious damage.

The average accident occurs at an impact speed of 15 kph, and tests carried out by the BIA at Thatcham in Berkshire use this speed as their test mode.

With care it is possible to reduce the repair cost factor to a very minimum. On average the repair bill in the UK is £480.00. This can be greatly reduced by moving items such as the battery away from the front of the engine compartment to the rear. A damaged battery must be replaced, at an average cost of £40.00, which is nearly 10 per cent of the total repair cost. The danger aspect of a broken battery need not be highlighted.

The radiator is another costly item that is often fitted to the very front of the vehicle, which means certain replacement in the event of an accident. This unit can be designed into a less vulnerable area, and certainly fitted away from the extreme front. A support cage can be designed and air ducted into the radiator, or it can be moved back and alongside the engine.

The rear of the vehicle must also come in for careful design. A well structured bumper assembly carrying impact shock into the chassis longitudinal will minimise panel damage to the rear of the vehicle, hence saving, perhaps, damage to rear light clusters. The position of the fuel tank and lines is of great importance, and ideally the fuel tank should be situated between the axles and underneath the vehicle. In the event of collision from any quarter the tank should then remain undamaged. Ruptured fuel tanks during collisions are not to be recommended and fitting them beneath the car will ensure that fumes do not enter the passenger compartment.

Fuel lines should be of metal tubing and should be run under the vehicle and fitted as close inboard as possible, giving adequate clearance to the exhaust system. The exhaust system can be so laid out that a rear collision will cause the very minimum of damage. It is preferable to swing the tail pipe across the rear of the vehicle so that rear impact merely bends the pipe rather than it acting as a lance right through the system to the engine manifold. A straight exhaust from front to back will certainly damage the whole system.

A well-placed spare wheel in the rear can afford considerable protection in the event of rear end accident. This can protect the interior of the boot and absorb some of the energy that would be expended into the passenger cell.

In the design of a vehicle it is necessary to think ahead, to plan *out* structure defects and plan *in* safety, usability and repairability.

Two comparatively new terms are increasing in significance and are fast becoming jargon. They are 'repairability' and 'damageability'. When designing for repairability, the meaning of this word has to be precisely understood. Good repairability applies when, for a measurable standard impact situation, the total repair cost is low. Low cost can either be due to low parts cost or low labour hours, or a combination of the two. A vehicle must be 'repairer friendly', providing good accessibility.

Good repairability does not equal good damageability. It is all very well

Design

claiming good and easy and cheap repair characteristics, but the question must be posed – should it have been damaged in the first place?

Good damageability implies that, for a given measurable impact situation, the extent of the damage is low. This necessitates good energy absorption characteristics, either by the progressive collapse of the structure, or by the use of energy absorption materials, or both.

The formula of space equals energy absorption equals low cost enters the design purpose for optimum damageability, and repairability need not and must not affect established safety or occupant protection criteria.

Above all things, safety of the occupants is paramount. Not only have the occupants of the vehicle to be placed in the safest travelling environment, the vehicle itself must be by definition as safe to other road users. Every item, from the passenger cell to the road wheels and tyres, must be bound by this safety consciousness. As already established, 65 per cent of impact is to the front of the vehicle, so seat belt anchorages, seat anchorages, steering wheel and column position and collapsibility must all be designed with the safety parameters in mind.

An area that leads to great concern is the interior fittings which protrude and can cause serious injury in an accident. Sun visors, interior mirrors, grab handles on fascias, and handles on glove pockets are all potential injury causing items. Soft foam padding should be much in evidence in the interior to minimise injury, as well as the fixing and wearing of rear seat belts. It is all very well restraining the driver and front passenger, but if rear passengers are not restrained they can cause dreadful injuries to themselves as well as to the occupants of the front seats. Legislation soon to be brought in will make the fitting of rear belts compulsory and it will then be mandatory to use them. Performance car manufacturers like Aston Martin Lagonda Ltd are well aware of this, and have been fitting inertia reel seat belts to both front and rear of all the models since 1985.

A well designed vehicle is one that is designed for safety; aesthetic good looks are a secondary feature.

When designing any structure there are many factors that must be taken into account. Not the least of these are the possible methods of manufacture by which the unit or component may be made. Some of the primary factors governing the design are as follows:

1. Initial design considerations:
   - Shape
   - Dimensions
   - Solid laminate
   - Sandwich construction
   - Polyurethane foam

2. Basic mechanical design:
   - Stiffness
   - Strength
   - Strain limitations
   - Weight
   - Means of support or attachment (metallic inclusions)

Vehicle Fabrications in GRP

3. Other design criteria:
   - Fatigue resistance
   - Creep resistance
   - Environmental resistance (moisture, chemicals, oils, etc.)
   - Temperature (low baking paint process)
   - Damage tolerance from impact or abrasion
   - Vibration damping (road shock)
   - Thermal expansion
   - Energy absorption

4. Processing criteria:
   - Dimensional tolerances
   - Quality
   - Subsequent finishing operations
   - Final assembly methods
   - Finishing

If the above can be incorporated into the thinking behind the design process then a solid foundation will have been laid to design and construct a vehicle that will be safe and have a useful life.

3
Mould Making

Every car manufacturer today has to construct a clay mock-up of the planned vehicle in full size. This enables the designers to see every line detail that has been drawn and it gives the stylists the opportunity to appreciate and rediscover their concept when the clay is complete and the decisions made. Then the process of turning this into the pattern for the steel presses commences. However, for vehicles that will be constructed in GRP the clay acts as a male plug.

The construction of the full size clay is a time-consuming procedure and it is important to persevere and ensure that the end result is as near perfect as possible. The moulding procedure is so quick, relatively speaking, that it can almost be discounted in the overall time consideration. Having committed the design to paper, the main construction can then begin. Usually, wood bulkheads are set up on a floor beam and key sections are fixed in position. Then a process of infilling takes place. When this is complete the longitudinal strips are put in. When the full frame is built then the panel infill can begin.

Construction of a full size clay

The primary construction of the frame is normally wood. A centre piece is set down and on this a centre line is drawn. Construction of the side supports can then be set at right angles and the side rails can be measured from the centre line (see Fig. 3, page 10). The bulkheads can be positioned and fixed accordingly, and then they can be supported to a rigid state. It is always desirable to construct a box-section for the rigidity that is required (see Fig. 4, page 11).

When this frame is fully built and well supported the process of fixing the longerons can begin. These give the full shape of the clay and begin to add the dimension of the vehicle.

Finally the infilling can commence, when light plywood or, in the case of flat sections, chipboard can be fixed over the whole frame. Normally a good quality wood adhesive, as well as mechanical fixing, is used at this stage. The complete clay must be solid with no apertures left. The doors, windscreen and sidelights must be completely solid.

Careful dimensional checks must be made at every stage and any alterations must be completed before proceeding. Any fault that is left at this stage becomes very difficult to correct at a later point. Care is the watchword here and the procedure must 'take as much time as it takes'. There is no quick answer to this time-consuming construction process.

When the point is reached where the decision is made to go no further, then claying up can commence. Clay is applied to the wooden substrate and is worked into the wood to obtain a holding key. As progression takes place a spatula should be used to commence forming the lines and working up the detail. This process must go on until the vehicle is a total clay. Then the curvatures and bodylines have to be carefully checked and measured out. This

Vehicle Fabrications in GRP

is a long and time-consuming operation but when correctly done pays dividends. A great deal of skill is needed at this juncture and it is something that cannot be taught overnight.

It is important to persevere with this; very good results can be obtained if great care is taken. It is essential that body curve pieces are constructed so that the one piece can be checked side to side on the body, ensuring conformity.

When the clay is shaped to satisfaction then any alterations that may be deemed necessary can take place. At this stage careful consideration must be given to the final outcome. Now is the time to alter bodylines, wheel arches, etc., because the final commitment will not lend itself to an easy alteration. Having reached this point, the final surfacing of the clay can take place. It can be smoothed down using wet and dry rubbing paper. The use of a block will help take out any undulations and assist in giving a very smooth surface.

Alternatives to clay

It is not necessary to work exclusively in clay to produce a mould plug. It is quite possible, and indeed desirable, for a less skilled person to move to another medium. Glass reinforced polyester in light sheet form can be used where a construction is more of a box shape, such as a cross-country vehicle or Range Rover type. The basic process is similar to clay in that a full wooden base section is constructed as outlined previously, but instead of clothing the structure in ply or chipboard two or three laminations of $1\frac{1}{2}$ oz mat can be laid on a flat sheet of contiboard and then cut to shape. After the full cure out has taken place it can be removed and fixed to the frame by mechanical means or by the use of deep filler as an adhesive. The catalyst fillers that are generally used in the automotive repair industry are ideal. They form well, and set very quickly and the adhesion is extremely good.

The whole plug can be constructed piecemeal in this way. At the end a skim of body filler can be applied at joints and fillet areas. When this has cured out it can be full sanded using 80 grit production paper. Thin smears can be applied and the plug shape worked up rapidly to a very acceptable standard. This format can be easily painted with an SR surfacer and wet flatted down to give a very fine finish indeed. A first-class mould can be taken from this structure.

Wood former construction

An alternative method for construction of a plug is the total use of wood. Again the base structure must be built up and the primary structure finished out. Then plywood can be used throughout the cosmetic stage. This is a little more difficult to work with as the sanding and finishing has to be to a very high standard and infilling can be difficult. However, for people with great experience in wood it may offer an ideal and practical way to achieve a plug for moulding.

Moulding returns

For a complete monocoque body the mould made for the vehicle will be a one-

22

Mould Making

piece construction. The use of separate panels in GRP and the moulding of these are dealt with later.

Naturally, once a mould has been made from a plug it will need to be undercut, or recessed, where door and window apertures occur. This type of composite, or re-entry mould, must be manufactured from the first pilot body piece.

Referring back to the design drawings, the door and window apertures must be carefully marked out on the GRP bodyshell. This operation can be built into the original clay or body plug, but often a situation arises where pieces will not fit to a tolerance that is desirable.

To complete this operation accurately, a GRP mould must be taken from the plug in one piece. This must then be braced with wood or angle Dexion for strength and support and then placed on a total support frame and fixed.

The mould must be left to cure out, normally for 4 to 5 days at a temperature of 68°F to 72°F, and then polished with wax release agent prior to laying up.

Returns and recesses can be fabricated up and fixed into position as a separate operation but for strength and ease of work it is preferable to lay up the total unit using split or composite moulds.

The only way to mould a re-entry or recess is to utilise a split mould. This is where the main body mould is over-fixed by a secondary mould carrying the shape into the interior of the bodyline. A typical door shut re-entry mould is shown in Fig. 17, which is a section through the door of an Estate car. The moulds are assembled before laying up and where the join occurs a fillet of soft wax can be applied (Fig. 18). Ensuring that a good quantity of gel coat is laid into this area and a fine tissue laid on that before the full laminations take place will guarantee a good finish to the area. The secondary mould has to be released before the main body mould. Normally it is held in position by mechanical means, and usually a nut and bolt fixed through steel brackets bonded to the mould is quite sufficient.

Split moulds

On large and somewhat difficult shapes the possible use of a split mould must be considered. The major difficulty with using split moulds comes in the accurate lining up and maintaining of conformity over a long period of use. GRP does tend to move and is affected by temperature changes. However, if the mould is extremely well braced and kept in a standard ambient temperature then these problems are minimised.

The decision of where to split a mould is often dictated by a convenient body line or panel change line, as well as the easiest release from the pattern, and subsequently the moulding from the mould. It is important to study the shell closely so as to obtain the best join line possible. Normally moulds are split into top and bottom half pieces. Lotus have used this excellent split on all of their models with great success, hiding the actual join beneath a trim piece.

Polyester resin coating

Once the clay mould has been completed it must be sealed back to take the gel

23

Vehicle Fabrications in GRP

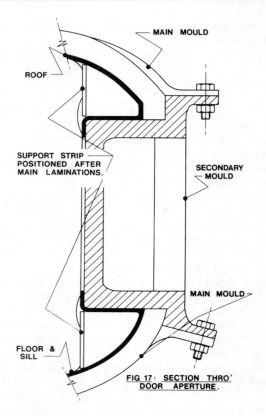

Fig. 17 Section through door aperture.

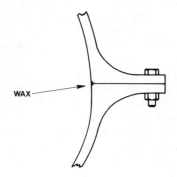

Fig. 18 Wax insert to stop gel coat penetration.

Mould Making

coat and laminations of the female mould. Polyester resin is widely used as a base. This material can be applied by spray or brush and must be carefully guide-coated and rubbed down when fully cured. Great care must always be taken to ensure that the finish is as smooth as possible and has as few nibs or marks as possible. Everything will show up in the finished mould.

Preparing the plug

When the finish of the plug is complete and it has a first-class moulding surface then the application of wax release agent can commence. It is desirable that this operation be left until the mould polyester coating has fully cured out.

Applications of wax release agent must follow the manufacturer's guidelines strictly; usually four or five applications are recommended. On the first pull it is advisable to use a PVA (polyvinyl alcohol) solution to ensure a clean lift. Always use soft mutton cloth for polishing so as to give a perfect lustre to the plug.

Taking a mould

After the application of the PVA, check that it has dried out completely and prepare the gel coat. Normally it is good practice to colour the gel coat with black pigment. This will always show up an area of low density when moulding from the female.

Apply two gel coats ensuring that the first sets up and that you do not pull up or damage the first application. Follow on with the chopped strand mat of the required weight ensuring that no more than three laminations are applied at any one time. The exothermic heat developed can be substantial and burning can take place. After six laminations the inclusion of support bars and brackets can be considered. This can be carried out only after the laminate is fully cured, to avoid surface marking. The use of Dexion, or slotted angle, is to be recommended as it is very strong and supportive, as well as having the holes for resin to migrate between layers of mat, ensuring a very positive bond and fixing (see Fig. 19).

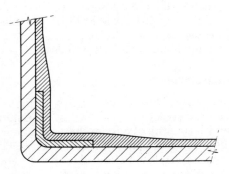

Fig. 19 Dexion strip located in GRP and then over-laminated.

Vehicle Fabrications in GRP

The inclusion of small timbers in the construction will help brace and fix the mould. It is necessary to build either Dexion or wood formers into a box-section around the mould, as in Fig. 20, so that the mould, after release, can be turned upside down for hand lay-up. Support bars of Dexion must be fixed across the top of the mould to give support and complete the mould into a box. These span bars must be removable so that the shell can be lifted from the mould.

When the total lay-up of the plug is complete it must be left for seven to ten days to fully cure out. After this period of time has elapsed the removal of the female from the plug can take place. There are various methods of easing the fixture, but generally it comes down to gentle movement and leverage applied in the right place to get a start, and then to force the female away. The introduction of warm water between the female and the plug helps break down the PVA and ease the release. Sometimes a gentle tap will start the release, but there is an ever-present danger of damaging the mould if this is carried out to too great an extent.

When the female is at last ready, then it can be prepared for the first male shell to be taken from it. Before carrying out this operation, it is important to examine the mould carefully for any imperfections. If small nibs or marks do exist then they can be flatted out of the gel coat, or filled if necessary with body filler and then polished back with compound.

Reinforced moulds

It is a fact that the stronger and better reinforced the female mould is, the better. This directly affects the longevity of the mould and the number of shells that can be produced before major overhauls are necessary. It is important to keep the mould in good condition, and well polished. The extra bracing and strength that may be added will help keep the standard of the shells consistently high. Repeatability and total conformity are the keynotes to success.

Use factors

The number of units that are planned from one female mould will be the criterion for its construction. If it is proposed that only ten shells be constructed then it can be argued that as it will have less work to do, a less robust unit need be manufactured. However, if over 200 are planned, more serious consideration must be given to the mould structure and design.

Alternative reinforcements

As well as using timber and Dexion type angles, other secondary reinforcements can help make the structure rigid. Strand Glassfibre offer a material called Scoreboard which consists of a foam that will accept resins and solidify when sandwiched between laminations. This is extremely strong and can be placed in or on any curved shape and overlaid with resin and matting. Rovings can also be introduced, along with materials such as Kevlar. These add greatly to the overall strength of the unit.

Mould Making

Facilities for production

Many factors come into play when a full production facility is envisaged. The number of units to be produced in a twelve-month period must be targeted clearly, and the introduction of a second female must be considered. Using only one mould may cause problems if damage occurs, and costly down-time cannot be endured. If a large operation is considered then the plug and one female mould should be stored in a safe dry place away from the work premises. A fire can result in months of work being lost and a tremendous set-back in business terms.

The repair and maintenance of moulds is somewhat of a preoccupation in the motor industry and the spurious panel industry. Constant attention to detail is the prerequisite for success and consistency.

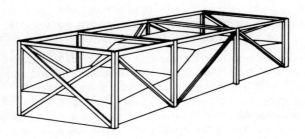

Fig. 20 A wood frame surrounding the female mould.

4

Taking Moulds from Vehicles

Over recent years the specialist car market has enjoyed some success in bringing to the public replica motor vehicles. Most of these are constructed in GRP and it is possible to purchase bodyshells of AC Cobras, E Type Jaguars, XK 120s and even Porsche 911s! These are normally sold as part of a DIY kit. This has all been accomplished by taking moulds from an existing motor car, which is quite easily done. As there is no copyright on vehicle design this practice is within the law. Developing on from that idea, the making of replacement panels, such as wings and bonnets for standard, rusty or damaged vehicles, is a natural commercial progression. These panels can be purchased from motorist shops and they have found a ready market. The rusting out of production vehicles is prevalent and will continue to be so.

To take a mould of a panel is quite straightforward. For example, a mould for a bonnet panel for a Ford Cortina MK IV can be produced as follows:

1. Remove the bonnet from the car and set it up on a firm stand.
2. Wash the panel with hot soapy water to remove all dirt and road film.
3. Check the panel for any dents, and repair if necessary.
4. Wet flat the bonnet using P600 wet and dry paper, and then apply three coats of SR grey surfacer. All the major paint manufacturers can supply a suitable material for spray application. After allowing an overnight dry through, wet flat the surfacer with P600 wet and dry for a fine finish.
5. Polish with wax release agent, taking care to polish the edges.
6. Polish the panel to the makers' recommendations using mutton cloth for a fine finish.
7. Apply PVA for the first pull.
8. Apply gel coat tinted black and lay up at least six layers of chopped strand matt. Roller well down.
9. Bond in steel support or wood to strengthen the female.
10. Leave for seven days and then release by introducing hot water between the bonnet and the GRP.
11. Allow to stand for several days after release, and trim female mould to remove excess and unwanted mat.
12. Polish the female in preparation for taking a panel.
13. Apply PVA and then apply gel coat. Follow on with three laminations of $1\frac{1}{2}$ oz mat reinforced with mild steel bar bonded in after the last lamination.
14. Bond in hinge plates and bonnet locking mechanism.
15. Release from mould.
16. Trim the return edges with a body file or block with 80 grit production paper.
17. Allow panel to stand or fit to the vehicle but in any case do not paint for at least seven days.

It is quite possible to prepare a mould from a complete vehicle, either panel by panel or in a whole piece. It is less complicated panel by panel but

Taking Moulds from Vehicles

nevertheless it can be conceived as a whole unit.

To take a mould from a complete vehicle in one piece demands careful thought and study. Observe where undercuts or re-entry panels occur, and bearing this in mind, decide where the split mould lines must occur to ensure a simple release.

To take a one-piece mould of an existing vehicle and then build up re-entry moulds using that piece as the plug is easier and preferable. As an example, consider taking a mould of an open sports car, such as an MG B. The sequence is as follows:

1. Remove all brightwork, door handles, bumpers, lights and windscreen.
2. Examine the body and repair any small dents with body filler.
3. Wash the vehicle down in hot soapy water and wet flat with P600 wet and dry paper. Dry off.
4. Degrease with a spirit wipe.
5. Mask up, and spray paint the entire vehicle with ICI SR grey surfacer P540-87 thinned with 33 thinners, 4 parts primer to 1 part of thinners. Apply three coats, leaving 30 minutes between each coat. Allow at least 16 hours before guide coating and wet flatting with P600 wet and dry.
6. Wash down the vehicle and dry off with a leather.
7. Commence polishing with recommended wax release agent.
8. Between dry-off periods stop up the apertures around the doors, boot and bonnet with Plasticine. Ensure that this is used as a sealer and keep it below the cosmetic panel line (Fig. 21). Apply PVA to ensure a clean release.
9. Apply gel coat tinted black. Ensure an even and high build all over the vehicle.
10. Lay up three laminations of 1½ oz mat. Allow to cure and repeat the process building in reinforcement bars, wood and Dexion type angle.
11. Allow seven days to cure out in ambient temperatures of 68°F to 72°F.
12. Use warm water to assist the mould release.

Barrier films

Although it is preferable to paint a steel body with an SR surfacer as described previously, it is not absolutely necessary. Other barriers can be applied to a body which will allow a good release, although there are drawbacks to some of these types:

1. A fine sheet of aluminium foil can be fixed on a panel using a spray adhesive. This has to be carefully done to ensure that no air pockets or wrinkles occur. Whatever is present at the time of moulding will be faithfully reproduced in

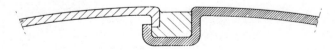

Fig. 21 The use of Plasticine to seal up the aperture between bonnet and front wing prior to mould taking.

29

Vehicle Fabrications in GRP

the final mould. After the application of the foil, a light polish with wax release agent will ensure that a pull, or fabrication removal, is satisfactory. Any imperfections in the mould must be filled with body filler and flatted down before the polishing process with wax release agent.
2. A good application of wax release agent, followed, after buffing, with PVA, will work well over original finish paintwork. However, certain vehicle refinish materials, for example cellulose, tend to be attacked by the resins and will break down in quite a dramatic way, leading to paint inclusions in the gel coat.

Construction of mould walls

It is essential when constructing split moulds that a mould wall or return is made. For external splitting, the simplest form is by using a fixed barrier, such as waxed contiboard, hardboard or rigid plastic (see Fig. 22). The barrier or flange is fixed in position and during lay-up the laminate is brought up to the flange as each successive layer of reinforcement is added. When it is fully cured the barrier or flange is removed. The surface of the mould exposed by the removal of the flange is then waxed and coated with PVA. The second section is then laid up in the same manner as the first. When fully cured the mould can be removed. Before removal drill and bolt the fixing so that it can be relocated.

Always bond in a steel plate where the drilling is planned, in order that the GRP mould does not fray in use. When mould walls are being constructed

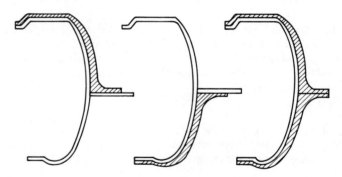

Fig. 22 Construction of a two-piece mould of a front wing.

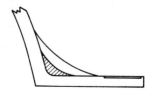

Fig. 23 Extra reinforcement.

Taking Moulds from Vehicles

ensure that a good depth of GRP extends out from the face, allowing ease of handling and fixing (see Fig. 23). It is important that moulds are bolted together securely, and room to work and manoeuvre a socket or spanner is paramount.

Always give a good build up of GRP at the change of direction as this will enhance the strength considerably (Fig. 23).

When a mould is bolted together the application of a wax fillet along the mould line will ensure that resin does not stray down the joint and cause release problems (see Fig. 18). Easy release means that the life of the mould is greatly enhanced.

Moulding complex returns

For an inner, simple shape a composite mould needs to be constructed. For more complex shapes it is the same formula, but using more complex composite moulds.

An inner return can be moulded by firstly moulding up to the edge of the return. When the laminate is in a 'toffee' state the edge can be carefully trimmed with a Stanley knife. When this has cured out a barrier of aluminium foil must be placed around the aperture. GRP can then be laminated into the recessed area. As this cures, steel fixing brackets, pre-bolted together, can be placed and laminated into the mould. When it is totally cured it can be unbolted and removed.

It is good practice to break the mould down and very carefully clean up the edges using a production paper. When it is necessary to manufacture box-sections for door pillars then a mould can be added after a previous lay-up and prior to removal of the complete unit. In Fig. 24 the outer wing of the vehicle is laid up with three laminations. This is allowed to cure out and then the secondary mould is bolted into position. A further three laminations are laid, interlocking the door structure with the wing. This gives strength and rigidity to the whole structure as well as ensuring that the section is sealed against water ingression. It is important to ensure that no sharp edges are included in a mould as GRP mat will not happily return on an edge and there is a tendency

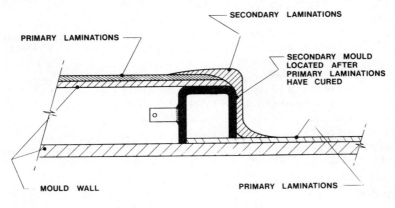

Fig. 24 Plan view of door pillar at A-post.

Vehicle Fabrications in GRP

for the resins to drain away. They also weaken the structure and should therefore be avoided.

Release agents

The release from the mould is totally reliant on the wax release agent used and its performance. There are a number of wax release materials, some applied as a polish and others sprayed onto the mould surface. It is important to follow manufacturers' recommendations for the best results, and as an extra precaution the use of PVA (polyvinyl alcohol solution) is normally recommended, at least for the first pull. Dry out times between applications of polish are important and for the best results, do take the necessary time. Poor release usually means a damaged mould, which can be time-consuming and costly to repair.

Gel coat

The gel coat is a thixotropic version of the laminating resin. This enables a mould surface to be coated by brush or spray to an even depth to give a robust cosmetic surface to the GRP structure. The gel coat can be colour pigmented with a whole range of colours including metallic, for a finish that will not break away from the substrate. This enhances the total appearance of the item constructed in GRP laminate. A double application of gel coat is advisable on moulds where there needs to be a margin before breaking into the mat itself. It is advisable to tint the gel coat black for moulds so that areas of poor opacity can be identified during the lay-up.

Mat and resin application

After the gel coat application the total lay-up of the mat can take place. It is advisable to lay not more than three laminations at a time as exothermic temperature can be induced by the catalyst. After application of lay-up resin over the gel coat by brush, lay in the first sheet of mat. Allow several minutes to elapse so that the resin takes up into the mat. Apply pressure with the roller to achieve impregnation. Apply a second coat of resin by brush, and apply the second sheet of mat. Then use the roller to achieve impregnation. Lay in the next sheet of mat and follow with the last coat and begin working the piece with the roller. Eventually a complete wet out will occur, and the random strands of the mat will be lying down but not broken up.

If further applications of mat and resin are necessary then ensure that the first three have cured out and there is no danger of exothermic reaction.

Lay down a coat of resin and continue as previously described for the next three laminations. Normally reinforcement can be built into the structure at this time and will give strength to the complete piece.

Cure out

Normally a full cure out takes three or four days. This includes the green

Taking Moulds from Vehicles

period when certain movements may take place. This period of cure out can be accelerated by use of a low bake oven unit. It is necessary to hold a temperature at the panel of 80°C for up to three hours to ensure a cure out. Normally GRP laminations are left to cure out at an ambient temperature of around 68°F to 72°F (20°C–21°C). During this time, as the catalyst reacts with the resin the GRP goes through a soft or 'green' stage. At this point it can be cut cleanly with a sharp knife, and this can be a great advantage when trimming the mould edges of unwanted material. Care must be taken not to cut into the mould or damage the face when carrying out this operation. It can save time, and a great deal of dust.

Before laying up, the glass mat should be cut by template to ensure the exact shape is accurate and repeatable during the lay-up. Small pieces applied at random will certainly lead to exothermic activity as well as uneven wall thickness of laminate. For a vehicle body, three lay-ups of 1½ oz mat is the appropriate weight for the structure.

Force dry and heat application

As stated previously, it is possible to increase the cure rate by the use of heat. A low bake oven installation can obtain the necessary cure in a two to three hour period, once the oven has run up to a panel temperature of 80°C. Other forms of heat application can be used but care must be taken to ensure the minimisation of the fire hazard (refer to Chapter 15 on Safety).

Infrared lamps can be suspended over the mould to radiate heat, which will enhance the cure rate significantly. However, it is important to understand precisely the power of this heat source as it has a very high temperature build rate. This can start moderately warm but within a short space of time be well above the desired 80°C. This form of heating is also affected by colours. Black, for instance, has a very high absorption rate and will heat up very quickly compared to white, which has a very much lower rate. It is advisable therefore to use this type of heating as a background heat source rather than direct radiation unless there is expert advice on hand.

GRP laminates are affected by temperature and therefore it is better to attempt construction during the summer if a fully heated workshop is not available.

Removal of the mould piece

After the cure out time has elapsed then the removal of the piece from the mould can take place. It is advisable to use warm water to break down the PVA and then to select one corner of the piece and attempt to break it away from the mould. This start is necessary to get a movement which can be worked on. The use of a small hammer to tap the inside of the piece to spring the laminate away from the mould may be necessary, but is not desirable on a repeated basis as it can damage the mould face. Normally, once a firm hold has been taken of the piece it should release quite easily. There are no rules for this except to state that if the mould has been correctly made and the release agent, along with PVA, have been correctly applied an easy pull should result. After the piece has

Vehicle Fabrications in GRP

been removed it can be inspected and replaced in the mould. This action will ensure that the piece remains held absolutely in place until it is necessary to proceed with the next operation.

Green time

This is the period when the GRP laminate has cured but changes are still taking place. In normal temperatures (68°F to 72°F) this lasts up to seven days. During this time any further application of materials for reinforcement, etc., will integrate very well indeed, and although more materials can be applied very successfully at any time, this stage is of paramount benefit. The bodyshell should not be painted during this green period as the styrenes present in GRP will tend to affect paint films and their long-term weatherability and adhesion standards.

Distortion

Forces at work in the GRP section make it vulnerable to distortion and it is essential to leave the piece in the mould for as long as possible so as to eliminate the distortion factor. However, it is possible to build reinforcement of steel or scoreboard into the construction to eliminate movement within the mould. It is important to lay up all the reinforcement pieces when the GRP film is still in the wet stage so that random movement can be taken up because of the non-resistance state of the film. Minor distortions can be corrected by clamping and reinforcement of the area and re-lamination, but as the distortion factor grows it is necessary to re-think the problem and the reasons why this has occurred. It is unlikely that any distortion will occur with three laminations of $1\frac{1}{2}$ oz mat on relatively small and reinforced areas on a vehicle body.

5
Preparing Moulds

Facilities

Planning of the workshop facility is important before commencing work. An area that is to be designated should have at least four feet working area around the plug or mould for safety and ease of operation. A cutting bench should be within easy working reach and all resins and catalysts should be stored in safe, purpose-made areas (see Chapter 15 on Safety). Lighting and heating need to be considered, and the appropriate action taken.

If full production runs are planned then a carefully designed workshop should be laid out. The number of units planned per annum, plus storage of both finished and raw material items must be carefully calculated. Expert advice is necessary here, and most production specialists can give advice.

Dry-out period

When the mould has been released from the plug then a period of time must elapse before putting the mould into operation, ensuring consistently good pulls. Normally seven days should elapse before polishing of the mould with wax release agent. The polishing cycle should be completed within two days and the mould will be ready for the first lay-up on the third day. Before the application of PVA, ensure that the mould is dust-free as any surface inclusions will damage the mould surface and contaminate the cosmetic finish of the shell. Any damaged edges or areas of the mould should be attended to at this stage. Filling of any imperfections should be carried out using catalyst fillers and prior to polishing.

Seams and edges

Where joining of split moulds takes place, the seam should be sealed to ensure that gel coat does not 'creep' along the surface of the joining area. To eliminate this problem wax or Plasticine can be used to infill the join area. Gel coat will go into this area that is slightly recessed but can be sanded off after release.

When producing a mould, the number of planned pulls must be considered. If this is a one-off operation then any slight defect can be counteracted when the shell has been released. However, if it is envisaged that in excess of 200 items will be produced from that mould, then the whole aspect changes quite dramatically. A normal production mould of this type will require careful setting up and a tremendous amount of reinforcement. Any movement during the moulding procedure will cause problems later in production. The life of the mould can be directly equated with the robustness of its construction and the way that shells will be released from it. Operator use and care will also determine the length of its useful life. The more rigid the structure the better,

Vehicle Fabrications in GRP

as a general rule. Remembering that the shell is green and therefore more flexible, it is easier to flex this out of the mould rather than the reverse.

Care of the mould cosmetic surface cannot be over-emphasised. It is important to keep it clean and dust-free at all times. The static that builds up must be treated with care and it is advisable to earth the mould if possible. Dust is attracted to the mould as a result of the static and must be carefully cleaned off.

When applying the wax release material use a mutton cloth and work in a circular motion. Start from one end of the mould and work methodically to the other. It is most important to ensure a uniform covering of wax and any area missed will cause the gel coat to adhere and then cause a breakdown in the mould as the shell is separated from it. Repair after such an event is very time-consuming and always unnecessary. It is to be avoided at all costs. Ensure that the wax is polished off and that a very high gloss finish is apparent. If necessary polish the mould again, and lift the gloss level as high as possible.

Mould assembly

If a split or composite mould has been constructed, then take great care in the assembly of the total mould. Bolt fixings must be carefully done up after the secondary mould has been placed accurately in position. If a discrepancy occurs here it is extremely difficult to remedy the situation later. Ensure that the mould piece contact is free of dirt or contamination and that the joining seam is filled with wax or Plasticine.

Mould rotation

With a complex or re-entry mould it is advisable if going for a production run to build the mould so that it may turn on an axis. This ensures that every lay-up position can always be down-hand. No matter how careful the operator, an amount of resin drain must take place. If it can be laid up down-hand, however, the right wet out takes place and all exothermic activity is kept to the very minimum. The mould needs to be constructed so that it may pivot longitudinally and can be locked preferably, in any position, but at worst case certainly a minimum of three positions either side.

When using a rotational mould take care before moving from any one position that the previous lay is sufficiently cured so that it does not drop away from the mould face.

It is advisable to check the surface of the mould, and the various fixings in the case of a split or composite mould, after every pull. This simple mould maintenance will ensure that, as the mould begins to show any wear or breakdown, remedial action can be taken. The care of moulds starts from the very first time the unit is used. A careful eye given to the quality of the shell and the female mould is a time-saving and money-saving action that pays dividends. A carefully maintained mould can last almost indefinitely, provided that the maintenance is carried out immediately and not left for days on end while the operator makes do with either some infill material or masking tape. It is a fact that once the surface commences to break up, unless this is attended

Preparing Moulds

to at that time the condition will worsen rapidly. In simple terms, give the mould care and attention and it will last for a considerable period of time and give consistently good bodyshells.

6
Laying up

Choice of mat weight

There is a complete range of materials available for construction. The most popular in vehicle construction is the 1½ oz (450 g/m²) This is heavy enough to give the required strength and stability to the body.

The three grades of chopped strand mat (1st grade) are:

300 g/m² – 1 oz per sq.ft
450 g/m² – 1½ oz per sq.ft
600 g/m² – 2 oz per sq.ft

Woven rovings are a heavy open weave fabric which is used for construction requiring high tensile strength with flexibility. This is supplied as follows:

280 g/m² – 8.3 oz per sq.yd
560 g/m² – 16.4 oz per sq. yd.
600 g/m² – 18 oz per sq.yd
830 g/m² – 24.5 oz per sq.yd

Chopped strands are glass fibre strands that may be mixed into a pre-catalysed resin to form a dough-like mix for infilling or bridging gaps. This is normally supplied in 12 mm strands.

Roving is a reel of glass fibre 'string' used in equipment designed for chopping glass fibre and spraying it in conjunction with resin and catalyst to produce a laminate. This material may also be used for winding applications and unidirectional reinforcement.

There is a range of polyester resins which can be used in the construction of a vehicle body. Those listed below are manufactured and supplied by Strand Glassfibre. They are as follows:

Resin A   A general purpose lay-up resin, top quality marine grade to BS and Lloyd's specifications. Suitable for vehicle construction, marine application and general work.

Resin B   A gel coat resin, marine grade to the same specification as Resin A. It is a thick resin which sets with a tacky surface on the exposed side and is used as the first resin coat in the mould.

Resin E   A flexible resin for producing a more resilient laminate.

Resin F1   For use when producing a laminate to comply with British Standard 476 Part 7 1971, Class 1 surface spread of flame, and British Standard 476 Part 6 Class 0 propagation test for materials.

Resin F2   For use when producing a laminate to comply with British Standard 476 Part 7, 1971, Class 2 surface spread of flame.

Resin H   A heat and chemical resistant resin. It will withstand heat up to 120°C.

The shelf life of resins is between three and nine months, and for maximum life they should be stored in a cool, dark place.

Laying up

The hardener used in polyester resins is in liquid form and is 50% MEKP (methyl ethyl ketone peroxide). An accelerator is available, which is 6% cobalt naphthanate solution. Some resins are pre-accelerated in production. Hardener (catalyst) and accelerator must not be mixed together as they will react explosively.

The quality of Strand Glassfibre resins is very high and the consistency of performance noteworthy. Suffice to say that these materials are used exclusively by Aston Martin Lagonda Ltd in the construction of air dams, spoilers and many small trim pieces contained within the vehicles (Figs 25, 26 and 27). These GRP laminates can be painted to a very high standard (Fig. 28) or trimmed in best quality leather (Figs 29, 30) and fitted to the vehicle (see Figs 31, 32, 33, 34, 35, and 36). For safety purposes Aston Martin Lagonda Ltd use the fire retardant to Class 1 specification resin.

Fig. 25 A driving scale model of an Aston Martin Volante with a body constructed in GRP.

Fig. 26 An Aston Martin Vantage air dam constructed in GRP and fitted prior to painting in body colour.

Vehicle Fabrications in GRP

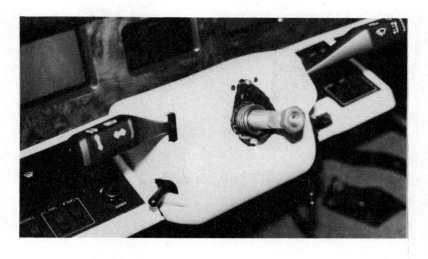

Fig. 27 (a) Mould for the Lagonda steering cowl. (b) The finished GRP piece trimmed in leather and fitted.

Laying up

Fig. 28 Painted air dam.

Fig. 29 Aston Martin V8 rear seats fabricated in GRP and padded and trimmed in top quality leather hide.

Vehicle Fabrications in GRP

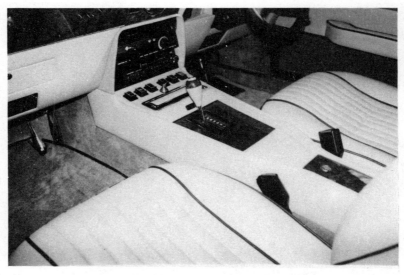

Fig. 30 Aston Martin console fabricated in GRP and trimmed in top quality leather hide.

Fig. 31 Aston Martin V8 transmission tunnel fabricated in GRP.

Laying up

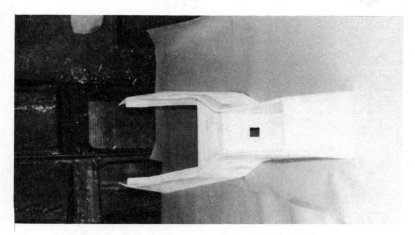

Fig. 32   Aston Martin centre console before trimming.

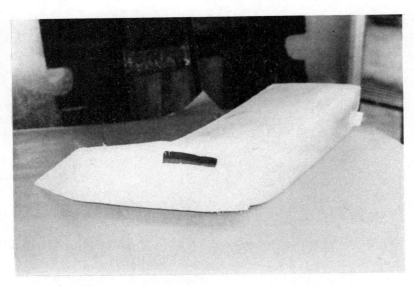

Fig. 33   Lagonda rear seat side panel in GRP.

Vehicle Fabrications in GRP

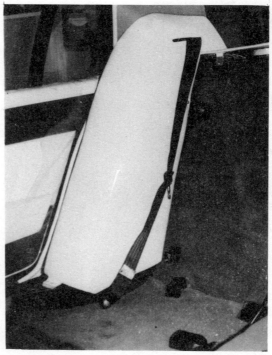

Fig. 34 Lagonda rear seat panel trimmed and fitted at first stage.

Fig. 35 (a) Lagonda rear seat side piece trimmed and fitted.

Laying up

Fig. 35 (b) Rear door panel.

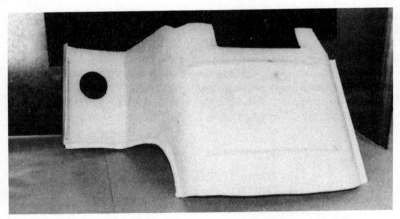

Fig. 36 Rear seat GRP piece for a V8 saloon before trimming.

Tissue mat in gel coat

In the construction of a mould, especially one that will be required to produce a large number of shells, an added strengthener can be admitted to the gel coat application. After coating the plug with the first gel coat and before a second application of gel resin, a single laminate of tissue can be draped over the

45

Vehicle Fabrications in GRP

mould. When the second application of gel coat is applied it will break down the tissue and this bonding will strengthen the total film considerably. The normal wear and tear on a mould surface does without doubt induce surface cracking and the application of this tissue is of great benefit.

Mat lay-up

When laying up the mat it is important to get total penetration of the resin and on vertical surfaces this is difficult due to the constant drainage that occurs. It is important to be able to lay resin and mat down-hand at all times. Large moulds that are used for mass-production are normally fitted up so that they may be rotated. If it is not possible to rotate the mould then application of resin must be worked up the side and after the mat has been laid in, the roller must work up the face of the piece to direct resin into the higher parts of the mould.

Constant working of the piece whilst the chemical reaction is taking place between the resin and mat is the sure way to obtain good penetration and wet out.

Observe carefully where resin drains down as areas of extreme exothermic reaction can occur in these places.

Pattern cutting

A pattern of the mat needs to be made, and normally hardboard serves extremely well. Always make the pattern as large as possible and where curves or re-entry occur, cut the mat accordingly (Fig.37). Overlapping will occur here and watch for any exothermic reaction, but the cutting will give a good drape to the mat and ensure a close lay-up without damaging the mat.

Before making a hardboard pattern, if doubt on dimensions exists, then produce a pattern in brown masking paper and check it against the mould. This will ensure greater accuracy before committing a hardboard piece.

Pattern cutting by hardboard is necessary when a long production is envisaged. For a one-off construction a paper pattern is quite sufficient.

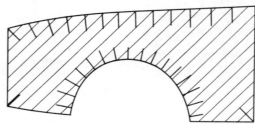

Fig. 37 1½ oz mat cut to template with further cuts made to ease lay-up into the mould.

Use of rollers

To produce a first-class laminate two areas of application must be understood. Firstly, a chemical reaction takes place between the resin, the catalyst and the

Laying up

mat. The end result is a complete wet out of the mat. However, it is necessary to induce this wet out by mechanical means and the roller action completes this bonding.

When mat is first laid down, a light roller action to bring it into contact with the resin underneath will allow migration of the resin up into the mat. If this is left for a few minutes then wet out will be easier and more quickly apparent. Full rolling will cause air bubbles to be forced out of the laminate and a good close contact film will be fully achieved.

Ensure rollering is as consistent as possible for pressure and roller in all directions. Use a small roller to gain access to the smallest area and to finish off over the top of edges. If it is desirable to have a fine finish on the inner surface from the mould face then tissue should be applied and gently rollered down to the laminate, allowing the resin to draw up through the film

Reinforcements

For vehicle construction there are a number of reinforcement materials that need to be considered. GRP laminates are extremely strong and only minimum reinforcement is necessary to obtain the results required. Four materials can be considered here and one of them will certainly give the required specification for the reinforcement of a vehicle body.

Steel

Either in bar form or 'Dexion' angle, this material provides an enormous strength factor when bonded into a laminate form. Mild steel bar bonded into windscreen uprights will give extra support to the roof, minimise movement around the windscreen area, and strengthen the passenger cell (see Fig. 7, page 13).

'Dexion' type angle can give support to boot floor area where a spare wheel may be carried, or as structure support underneath the floor pan. It can be used in doors or door pillar areas and, when bonded in at right angles between floor and side panel, is incredibly strong (see Fig. 19, page 25).

Fig. 38 A Kevlar reinforcement strip laminated into a wing edge.

Vehicle Fabrications in GRP

Kevlar

This material, supplied in ribbon or sheet form, is extremely strong when laminated in. It is ideal for further reinforcement of the windscreen area, and can be used in door and wing edges (Fig. 38).

GRP section

When laying up boot, bonnet or doors, the added strength on flat panels of a GRP section can usefully offer lightweight strength. A small drape mould can be easily constructed (Fig. 39) and a two or three lamination of GRP formed. After removal and trimming it can be laid in any appropriate place and then over-laminated (Fig. 40). Any size of GRP section can be constructed giving great versatility to this inexpensive form of reinforcement.

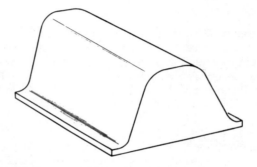

Fig. 39 A drape mould for reinforcement section.

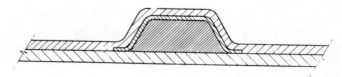

Fig. 40 Section of laminate reinforcement placed in position and overlaid. Centre section filled with polyurethane foam for extra rigidity.

Scoreboard

This material is supplied in sheet form and can be placed at any point within the vehicle. It is cut in such a fashion that it will follow mild body contours. It will take up resin and become impregnated, giving a solid structure within the laminate.

7
Reinforcement Process

The reinforcement process aims to structure the vehicle body in such a way that with minimal weight gain a robust structure, fully capable of carrying out the design functions and offering the maximum possible safety margin to the passengers, will be achieved. It is quite possible to design in a reinforcement that can be detrimental in the case of an impact accident. Careful design, revolving around the implosion collapse and absorbent energy factors, is important.

Reinforcing a structure at the front area leaving the engine compartment bulkhead in a weaker state is likely, on frontal impact, to take the front section into the passenger cell. A similar case exists for rear impact and the over reinforcement of the boot area.

Ideally, in the three compartment construction of engine, passenger and boot cell, the passenger should remain in a box form as the strongest structure with the two outer cells subservient to it.

It is preferable to construct the reinforcement within the passenger cell and then by degrees scale down the structure away from the cell to the extremities, reducing the reinforcement gradually. There can be no substitute for total passenger safety in that area.

The fact that in the design of the vehicles shown the floor pan and bulkhead are as a complete unit, forming a monocoque structure which is then fitted to a chassis, gives an immediate base line of boxed strength (Fig. 41).

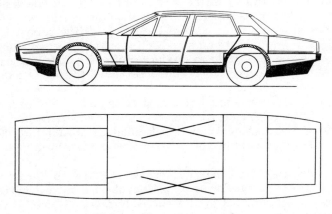

Fig. 41 Floor pan strength.

GRP box-sections

The use of box-section GRP has already been mentioned. This is a very inexpensive way of reinforcing a structure. By use of size and lamination levels

Vehicle Fabrications in GRP

a degree of accurate reinforcement can be bonded into the vehicle. Sections of channel can be decided upon, for example 1 inch × 1 inch, 1½ inch × 1½ inch and 2 inch × 2 inch could be designed up. The 1 inch could be two laminations of 1½ oz mat, the 1½ inch could be three laminations, and the 2 inch could be four laminations.

This section of channel bonded into the total film would give an ascending strength into the passenger cell. Support for door skins can be fashioned in this way, adding to the safety of side impact. The use of polyurethane foam as a support filler in box-sections after laminating into the main film is an excellent way to give further strength to the structure. Doors are panels that are an obvious choice for this extra support.

Lamination weight

A vehicle body constructed with three laminations of 1½ oz mat fitted to the chassis shown in Fig. 1 (see page 9), will give an approximate total vehicle weight of one metric tonne. If a Ford 1600 OHC engine is used, it will provide adequate power and performance for the vehicle. The Copperleaf Cars Ltd design of the Monaco is an example of this build, and it weighs 990 kg (see Fig. 2, page 10). For a one-piece body structure, three laminations with adequate reinforcement are quite sufficient for a construction of this size.

However, it is possible and quite feasible to carry on applying laminations to the shell until an enormous strength factor is built in. Normally, three laminations of 1½ oz mat with a gel coat will give a total laminate thickness of approximately 4 mm. In most cases this is quite adequate, but for every extra lamination a further 1 mm will be added to the thickness.

The use of rovings as reinforcement will give much greater strength to the structure, and in the case of 'off-road' vehicles this may well be desirable.

Because GRP laminates are more flexible than metal, stiffening by various means needs to be incorporated into the design of the panels. It is more advantageous to design in strength and stiffness to the former alone making the unit easier and less costly to produce. Stiffening added as a post-cure operation can cause a depression on the outer skin if the fitting is carried out before the moulding is fully cured.

Any form of stiffness should be aimed at spreading the loads over the greatest possible area. A most convenient and widely used method is to use a lightweight core of foam which is over-laminated with reinforcement. The foam acts as a former with the hollow section of laminate forming the rigidity to the moulding.

As most panels are curved, the reinforcement material must be able to follow contours where necessary. The use of plastic tube, balsa wood and cardboard cut to a 'top hat' shape can all give the curved former on which to overlay the laminate reinforcement.

To mould in a stiffening rib on a contoured panel:

1. Lay up the panel with gel coat and sufficient layers of mat and resin.
2. Before the laminate gels lay in the reinforcement former (foam, etc.).
3. If the laminate has cured, fix in the former by the use of an application of fast curing resin.

Reinforcement Process

4. Adhesives should be avoided as they can react with the resin and inhibit full cure.
5. Overlay the former with resin and mat.
6. Ensure that the position of the former gives an adequate area of laminate in contact with the panel each side.
7. Use a brush to wet up and roller carefully up to the former.

To give added strength and stiffness the following points are worthy of note.

1. Exterior styling lines
In the construction of modern pressed steel vehicles styling lines are used to strengthen the cosmetic panels of the vehicle. This exercise can be used with effect in GRP construction. A single deep concave flair line can be present throughout the length of the vehicle. This will give style as well as the strength associated with the curvature and change of direction in the GRP laminate. This exercise can be carried further to incorporate bumping strips protruding from the surface of the cosmetic panel, or a whole lower section of wing and door raised out to include the bumping strip. This will reduce considerably the bland area of a door and add considerably to its rigidity.

2. Wing to body areas
In certain styles the wing to body design can give complex double curvature sections that can increase and decrease throughout the wing to body moulding. Even lightweight laminates will become very strong where there is such a complex change of direction. In this case the mould would have to be split to enable such a structure to be moulded and released.

3. Double skinning of wheel arches
The inner wheel arch is usually moulded separately and then bonded and overlaid with laminate. By manufacturing the structure in this way great strength and rigidity is built into the final section. It is a case of double curvature panels being bonded together to form a many-dimensional supportive area. The contribution that this makes is considerable.

4. Roof panels
Roof panels are generally the largest unsupported panel in a motor vehicle. On Coupés or close 2 + 2 sports cars this is not such a problem. Headroom precludes the facility of deep ribbing; however, a deep curvature in the roof panel with the added strength of extra laminations can greatly increase the rigidity. Ribs or stiffeners can be incorporated into the areas above the front and rear screens, and along the side of the roof panel. Double skinning with the use of a foam sandwiched in the structure will provide a considerable level of strength and rigidity.

5. Bonnet and boot
Both bonnets and boot lids can be strengthened by the tasteful use of styling lines. If these are kept to the centre and mid-section of the panel, then a return of ½ inch with the addition of a rib set in round the outer boundary of the panel will ensure that the panel has considerable strength. The choice of stiffener necessary very much depends on the size and style of panels. A Mini Metro size

Vehicle Fabrications in GRP

bonnet will require less support than a MK 2 Ford Granada, for instance. It is therefore important to understand the design principles behind the structure size.

By using GRP and fully understanding its great strengths and its weaknesses, this experience and the learning curve that the new user develops, will produce a 'feel' for the material which will greatly aid in the successful design and construction of a vehicle in GRP. There is no substitute for using the material to gain the experience necessary to fully understand the implications of designing for strength.

8

Release from the Mould

This is an area that can cause a great deal of concern when first using GRP laminates. However, if the correct procedures are carefully followed throughout then removal does not present a problem.

Procedure

Bearing in mind the conditions in which the female mould has been laid up, i.e. the temperature day and night, decide how long the new shell should remain in the mould. Two to three days is ideal if the temperature has been in the 68°F to 72°F window. Start a corner or edge by using a blunt tool, such as a wooden strip. When there is sufficient gap, pour warm water into the aperture to break down the PVA. If possible, from this point flood the mould. If not, then start on other areas and repeat the procedure. This activity should release a one-piece mould with little difficulty.

Section removing

Where a split mould has been constructed, it is quite a straightforward operation to remove the section after unbolting and attempting to pour warm water through the piece to break down the PVA if it is not possible to lift the piece away immediately. Once the main structure is open, it is quite easy to ease the shell away from the mould and continue with the warm water treatment.

Lift the sections away and place in a safe area so that they do not suffer damage during the removal of the main body shell. Clean any resin or gel coat that may have ingressed during the lay-up as soon as possible.

Impact removal

Shells that on the first attempted removal will not yield, can be lightly impacted. Using a small leather head mallet, start at one end of the mould and tap the shell laminate: the surfaces should spring apart. This is not to be recommended, as surface gel coat damage may result, with stress areas raised by the continual impact. Check after every few feet of release to see if the shell can be moved from the mould. Often only a start is required.

If the mould is pigmented black as suggested earlier, then when the laminate springs free it will appear as a milky finish beneath the laminate, thus clearly showing where it has released to.

Compressed air

Some moulders use compressed air for ease of removal but air under pressure is dangerous and is not recommended unless expert advice is available.

Vehicle Fabrications in GRP

Damage areas

If, due to a difficult removal, mould damage is sustained, it is important to repair this as soon as possible. Ideally the area that is damaged should be cleaned of any release agent before the repair is attempted. Normally the damage is confined to stress lines in the cosmetic finish of the mould. These can be repaired by rubbing them back with wet and dry paper (320 or 360 grit) and carefully applying filler as a fine skim. Then, after curing, follow on with wet and dry paper and smooth the filler into the gel coat. Rewax thoroughly and continue with moulding.

Cleaning the shell

Once the shell has been removed from the mould, and prior to any work being carried out, it is sensible to wash the shell with hot soapy water to clean off as much of the wax release agent as possible, as the agent will contaminate all future operations. If filler is to be used, the presence of the agent is very harmful to the adhesion qualities of the filler.

By handling the shell, the release agent can be transferred to laying up of mat and materials. Also, the presence of wax will greatly affect any paint procedure planned. Apply hot water with a measure of a household liquid soap for maximum effect. Clean water wash down and dry off with a leather or compressed air gun.

Cleaning the mould

After a number of pressings have been taken it will be necessary to clean the mould. Each GRP manufacturer will be pleased to supply a suitable material that will harmonise with the materials in use. Strand Glassfibre recommend Mirrorglaze MGM1 Heavy Duty Mould Cleaner.

Preparation to relay

When a production run is planned then a mould will be in almost continual use. To be effective ensure that all mat is cut to pattern and is laid out so that it may be easily applied to the mould. Also ensure that gel coat mixing is very thorough and the right amount of materials have been drawn from the store to lay up in one attempt. Although one can return to a mould at any time and continue, it is better to complete the total lay-up in one attempt. This will ensure an even cure through and good and total wet out and link up. It will also ensure that dust is not entrapped between laminations.

9
Fixing Bodyshells

Types of fasteners

There is a whole range of bondable fasteners now available on the market. Figure 42 shows a selection of these that can be laminated into the GRP at the time of lay-up. However, these can be fitted at a later date when lining up panels and parts if it is not possible earlier in the construction. These fasteners are invaluable for lightweight fittings and for joining purposes.

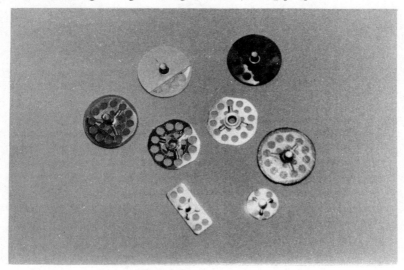

Fig. 42   A selection of the 'Bighead' fasteners.

Bonding of fixing points

At the point of attachment between the monocoque shell and the chassis, there should be a suitable bondage of the fixing point. A simple but effective way to strengthen an area is to bond in a plate of mild steel within the lamination which, if pre-drilled at the four corners, will provide a stable and well-bonded plate to finally drill and fix into position (Fig. 43).

It is important to ensure that the fixing point is not a stress raiser as this can lead to cracking and splitting. An area that is ideal for fixing is where panels are bonded in all directions, such as the footwell by the engine bulkhead.

For lighter fittings, washers can be used bonded into the laminate, and then drilled through after curing. This supports the bolt and stops fraying of fixing holes.

Vehicle Fabrications in GRP

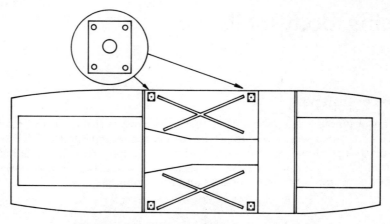

Fig. 43 Steel plate reinforcement in floor pan.

It is important to ensure that the fixing points are as accurately positioned as possible to allow for chassis fixing with the minimum of stress. To tighten a monocoque shell down when there are areas of even minor stress will eventually lead to fracture. It is advisable to fit a rubber pad between the fixing point in the body and the chassis (Fig. 44).

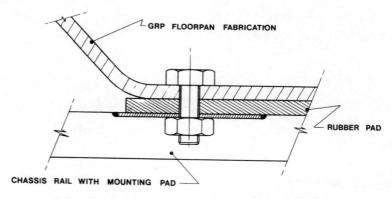

Fig. 44 Insulation between GRP and steel chassis.

Reinforcement

At the fixing point it may be necessary to spread a considerable load at major points. This can be done by fabricating steel plate into two or more planes and laminating the piece into a heavy lamination section (see Fig. 8, page 14).

The use of mild steel box-section 1 inch × 1 inch with a plate welded to it, and then the whole piece laminated into the structure, will give great torsional strength (Fig. 45).

Fixing Bodyshells

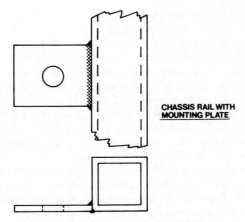

Fig. 45 Box-section with welded plate.

Strength of areas

The main area for chassis fixing must be the engine bulkhead and the floor pan. These panels must be four or five laminations thick with the necessary reinforcement around the fixing areas. Some items of in-vehicle equipment will bolt through the floor to the chassis giving a sandwich effect. Seat belt anchorages and handbrake brackets are two examples. It is important to fix pedal boxes and steering gear to steel chassis mounted brackets to ensure stable and safe fixing to the structure. The GRP is ideally sandwiched by this fixing arrangement (Fig. 46).

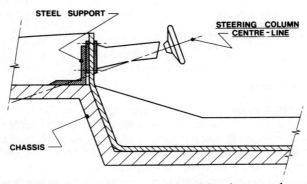

Fig. 46 Pedal box mounting through bulkhead into steel support.

Decisions on stress points.

Taking the main floor pan as the area where 90 per cent of the chassis fixing is to occur, it is important to spread the fixings over the greatest possible area

Vehicle Fabrications in GRP

(Fig. 47). As well as major fixings in the main structure, great strength can be obtained by fixing at the extreme points.

Stress may build up in the shell as a result of fixing, and this can be relieved somewhat by the use of a cushion material. Bolting through to the chassis using rubber pads is a major step in that direction but this can be carried further to include a complete buffer of rubber strip between the shell and chassis.

The Monaco, designed by Copperleaf Cars Ltd, employed this method and eliminated a high percentage of road shock and vibration from the monocoque bodyshell. This made the vehicle comfortable and reduced noise considerably.

Fixing the body where it 'lays' is a certain method of reducing the stress that can be imposed when bolting down. There is always likely to be some movement in the GRP as it fully cures out and this may take weeks or even longer. To overcome this problem, fixing and pulling down on the main fixings first will allow the floor pan to sit on the chassis and where extreme fixing points occur it may be that the pan is not firm down on the fixing pad. At this point use GRP sheet as packing to build up the area, before bolting through. This ensures that no stress point is raised by tensioning the GRP when tightening the bolt.

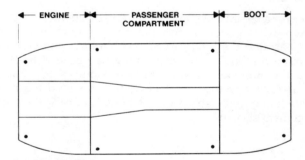

Fig. 47   Floor pan showing extreme chassis fixing points.

Easy shell removal

It may be that after a vehicle has been involved in an accident it is necessary to remove the shell in order to repair chassis damage. It is therefore important to bear in mind when designing the vehicle the ease with which the body may be removed. Normally, with the vehicles used in the examples, the body may be unbolted and lifted off after the disconnection of the various mechanical items. It is advisable to use free bolts that can be completely removed from the structure rather than bonded-in units. A degree of sideways or backwards movement is always necessary to remove a shell. To lift off upward with no difficulty is uncertain when bolts are bonded and the front and rear underpanels may foul the chassis or support bars.

Once the chassis is free from the body then a safe and expedient removal is to raise the body clear enough for the chassis to be run out from underneath,

Fixing Bodyshells

leaving the shell on stands clear of the chassis run. If the fixing points are accessible without disturbing much of the inside, it keeps costs down. If the vehicle has been involved in an accident that has damaged the chassis then body damage will also have occurred, and therefore if the shell is safely positioned on stands it can be worked on in that position.

Bolt tension

When fixing the body the use of high tensile bolts is necessary and the nuts should be of the self-locking Nylock type. This ensures that no fixing will come loose through body and road vibration. The bolt should be drawn up so that it just begins to compress the rubber pad and no more. This is quite sufficient and will ensure that no unnecessary tension is placed on the shell.

Lubricants

When bolting down the shell to the chassis it is recommended that a lubricant is used on the bolts. It is good practice and will ensure easy removal at any stage in the life of the vehicle. There is a whole range of chassis greases and easy-slip materials now available on the market. Protect the fixings as much as possible from the harsh effects of salt and road film that have such dreadful consequences on steel.

10
Trimming and Faults

At door, window, boot and bonnet apertures there will normally be a return of some description. On a continual production process it is necessary to trim these areas accurately and with consistency. This can be accomplished by the addition of mould lines. These can be carefully drawn into the plug so that when the female is created the line will be represented each time a shell is produced. This gives an accurate moulded line to trim to on every body. The depth of this trip line has to be carefully measured in the original plug to ensure that rubber mouldings or finishers fit and seat correctly on the vehicle. Obviously, around door apertures this makes the difference between good and wind-proof door closings, and water ingression.

If a return of 1 inch is required then it is a straightforward operation to develop a 1 inch block and fix a scriber to it. Chase round the apertures with this device and mark the line out clearly. Continue working round the aperture in this manner until a deep scoreline is visible and can be felt. Very lightly follow round with 80 grit paper folded to enlarge the scoreline and soften the edge. Ensure that wax release agent is applied and worked into the scoreline. This line will then give the operator a clear indication of where to trim.

Cutting and grinding

Firstly, always ensure a dust face mask is used during this operation as any GRP in dust form is harmful.

As previously mentioned, a shell can be trimmed in some areas when in the 'toffee' stage with a sharp Stanley knife. This is not always possible and some trimming must take place when the shell is removed from the mould. There are a number of tools that are available for this operation. Hand tools, such as hacksaws, body files and wood blocks with 40 and 80 grit production paper, are invaluable and always have to be used for the final finishing. Power tools such as a jigger cutter, or wheel cutter, as well as a DA sander with Stick It discs of 80 grit and 100 grit, are essential if production numbers are to be planned. Small air grinders are available which can be very useful. It is a fact that GRP causes problems to these tools by the nature of its make-up. Saw blades and grinding wheels soon lose their cutting edge on this material. Make sure that the jigger cutter is of a robust industrial type as a small DIY unit will break down very quickly indeed..

When the shell has been removed from the mould then trimming can take place immediately. Where mould lines have been set, a jigger can trace round quickly and accurately. Always remain slightly on the plus side and finish off with 80 grit production paper with a wood block. This will give better control and allow a rounded finish to the return.

Where a split mould has been in operation the lines resulting from the mould

60

Trimming and Faults

split are best taken down by hand using production paper. This can be done with a block or literally by hand when following a contour.

Keep all tools well 'dusted' out and use a lubricating machine oil to prevent clogging of the internal parts.

Filing and refilling

It may be that after using a body file an overfiled situation develops, with heavy score marks in the gel coat. These can be quickly and easily refilled using a body filler. This is the quickest and most satisfactory way of repairing any surface damage to a bodyshell. The adhesion properties of these fillers are now excellent and they can be used without any difficulty.

Edge repairs

Wherever there is a sharp edge or change of plane in the body style, damage can occur. The use of deep fillers as previously stated is quite acceptable, but in production the less filler used the better, and wherever possible the more original gel coat which is left intact the better. This is particularly so on edges, and therefore if damage is extensive or poor moulding has caused major edge problems then a remould of that section should be carefully considered. A new moulded piece can be fixed into position and rebonded very easily during these early stages.

Measuring apertures

For production methods the best way of making quick checks during the production cycle is by the use of jibs. These can be constructed from GRP and edge bonded with steel. Wooden formers are also quite acceptable. Time spent on constructing these tools can be invaluable when checking for distortions.

Drop check

A simple way of checking for any distortion or movement that may have taken place within the monocoque body is to carry out a standard drop test check. This consists of suspending the shell above the floor and from various balanced points dropping a plumb line to the floor and marking the position. Six or eight points should be marked down on the floor and these should be joined by straight lines. They should all cross on a centre line (Fig. 48).

Dry flatting

When dry flatting the shell with production paper use 80 grit or even 100 grit paper to ensure that the gel coat is not broken down. Once the laminate is uncovered then deep catalyst filler must be applied to rebuild the surface. On rounded contours the hand-held paper will work very well. Flat areas may be blocked, but care must be taken not to break down the gel coat and weaken the cosmetic structure of the laminate.

Vehicle Fabrications in GRP

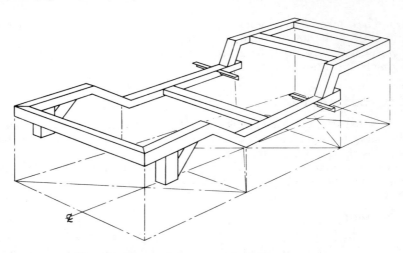

Fig. 48 Chassis drop test to check for alignment.

Preparing the gel coat surface

Careful examination of the gel coat will ascertain whether any air bubbles have remained near the surface during coating up. If these are detected then they must be broken so that they may be filled with catalyst body filler. Vehicle manufacturers will use a low bake oven facility to bake the shell and force any air bubbles to 'blow' out through the surface. As Lotus and TVR use a low bake painting facility it is imperative that any minor air pockets are attended to before the long process of high gloss finish is commenced.

Stress relief

In the event of an accident, the shell will absorb stress in the same way that induced stress may occur when overtightening any chassis fixing. This normally shows up as 'starring', and can be rectified.

The point where the stress has occurred should be examined carefully. The stress lines that normally show in the gel coat must be followed to their extremities. When this point is identified, a small hole should be drilled right through the laminate. This will stop the stress spreading throughout the laminate. The centre point should also be drilled through the laminate (Fig. 49). The holes that have been drilled can then be filled with catalyst body filler. This is a time-consuming operation but will be successful if done correctly, and will certainly save the shell from further deterioration.

When this operation has been carried out, two or three laminations of mat can be applied over the area from the underside. This completes the process and gives more reinforcement strength to the laminate.

Trimming and Faults

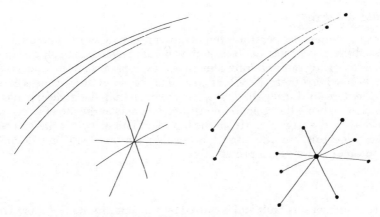

Fig. 49 Relieving stress in GRP fabrications by drilling.

Faults in GRP

Most of the problems that arise in a GRP lamination normally revolve around the resin. Undercure of the resin film is a major cause of a breakdown. There are other faults that occur, some of which are visible immediately and others which develop in time. Rectification of these faults is generally obvious when examining the fault. Care at every stage will usually eliminate the fault, as most materials now in use are very carefully controlled during manufacture.

Surface wrinkling

This is a fairly common fault and it is caused by solvent attack from the monomer in the laminating resin, and an undercure situation in the gel coat. Other conditions that can cause the fault are incorrect shop temperature or humidity levels, or the fact that the gel coat has been in a draught. Also, application of a second gel coat before the first has cured up can lead to softening and some interaction causing a surface wrinkling.

Pinholing

This is caused by small air bubbles which become trapped in the gel coat prior to its curing. The trouble begins if the gel coat is too viscous, has too high a filler content or if it fails to wet the release agent on the mould surface properly. Another reason for the fault is if the gel coat has been stirred too much prior to application to the mould face. This fault manifests itself when applying heat to the shell for either curing or painting procedures, as the air bursts out through the gel coat.

Vehicle Fabrications in GRP

Internal dry areas

This occurs when an operator attempts to impregnate more than one layer of mat at a time. It is vital to get thorough wet out at every stage of the laminate. Imperfect wetting out of reinforcement is only apparent on the underside of a lamination where there is no gel coat. The cause of the dryness is either insufficient resin during the lay-up or the fact that the rollering has been insufficient to bring the resin to the surface. Dryness at the top of a deep mould is due to resin drainage away from the top edge. Thixotropic resins can be used to overcome this but the use of a positional mould allowing down-hand application at all times is a preferred route.

Leaching

This is a very serious fault and is caused by exposure to the weather. There is a loss of resin from the laminate which leaves the glass fibre strands exposed to attack by the moisture in the atmosphere. The reason for this is that the resin has not been properly cured or it is the incorrect resin for the job in hand.

Spotting

This fault appears as small spots over the entire cosmetic surface and is caused by inadequate mixing or dispersion of one of the constituents of the gel coat.

Fibre pattern in gel coat

When the fibre pattern can be clearly seen near the surface of the gel coat it is due to the gel coat film being too thin. It can also happen if the laminate is removed too soon from the mould. When carrying out further work on the GRP it is important that the gel coat is not breached into the fibre layers.

Fish eye contamination

This fault occurs when a highly polished mould is in use and a silicone-based wax is used for the release. In this case the gel coat tends to move away at random from an area leaving a light film, and this shows as pale patches or even brush marks in the cosmetic film of the laminate.

Summary

If correct procedures are carefully maintained and followed consistently, all faults are avoidable. Conditions and operation will determine the fault factors in GRP moulding.

11

Painting a GRP Bodyshell

When the body shell is totally finished, a paint programme has to be considered. There are some schools of thought that advocate coloured gel coats. There are, however, many problems associated with this idea for final finish. For example, if the pigment is of poor opacity then patches will be present in the gel coat, giving a very poor appearance. Examine some kit cars carefully and this will be obvious to the naked eye.

It is important to understand that any small surface craters or holes cannot be filled with catalyst filler, as this will show. A further point to remember is that, should the vehicle be scratched or suffer minor impact damage, then it is impossible to remedy the situation. Generally the appearance of a pre-coloured gel coat is not as attractive nor as sophisticated as a full painted bodyshell. The choice of colour is limited in gel coat pigments, compared with 18,000 car colours on offer at present. The new system acrylic polyester clear over base is very attractive, and a superb paint finish will enhance the vehicle and change its appearance dramatically.

If the paint manufacturer's instructions are followed carefully then you can be sure of an excellent result. To begin with, the preparation must be careful and methodical. The shell must be completely free of wax release agent, and to achieve this the following procedure must be observed:

1. Wash the body down with a solution of very hot water and washing up liquid. The heat and the soap will soften the wax and help break it down. Repeat the operation several times.

Once the body is totally free of contaminate then the following paint procedure will ensure a first-class finish:

2. Wet flat the complete body with 360 wet and dry.
3. Dry the body completely.
4. Wipe over with solvent spirit wipe.
5. Mask up as necessary.
6. Spray one coat of ICI self-etch primer. This must be used for maximum adhesion. Leave 20 minutes to dry.
7. Spray three coats of ICI RPF 800 filler thinned with 396 thinners to approximately 40% filler to 60% thinner (25 to 29 secs BSB4 cup at 20°C).
8. Leave 20 minutes between coats and then overnight to through dry.
9. Spray a thin guide coat of ICI P030 black thinned with 396 thinner (approximately 80% thinner to 20% black) over the complete body. Do not try to obtain obliteration.
10. Wet flat the complete body using 600 wet and dry. Use a rubber or wooden block on flat surfaces. The guide coat will show both low areas and areas where flatting has not taken place.
11. Dry off completely.

Vehicle Fabrications in GRP

12. At this stage stop up as necessary. Any small marks or indentations can be filled with ICI cellulose stopper. Leàve at least three hours before attempting to rub down stopped areas. After rubbing down spray filler RPF 800 over the stopper to seal it in.
13. If the surface is smooth and there are no blemishes then wipe down with solvent spirit wipe and prepare to spray colour.
14. Straight colours: Mix up the colour thoroughly and thin with ICI 804 respray thinner, to approximately 60% thinner to 40% colour (24 to 28 secs BSB4 cup at 20°C). Spray four single coats over the vehicle. Leave 30 minutes between each coat, then leave overnight to through dry.
15. Wet flat the vehicle with 800 wet and dry. Ensure that the whole body is evenly flatted. Dry off and examine. Any minor indentations previously missed can be stopped up at this stage and then must be carefully rubbed down and respotted with primer. When this is dry it must be flatted with 800 wet and dry.
16. Solvent spirit wipe the vehicle.
17. Spray two single coats of ICI colour thinned as before, and leave 30 minutes between each coat.
18. Spray a wet on wet coat of colour to achieve maximum build and flow.
19. Allow at least seven days to through dry.
20. Polish with ICI compound 2A and repolish with Tetrosyl T Cut, and finally apply a light wax polish.
21. To achieve an even more outstanding finish, instead of polishing after seven days, wet flat the vehicle using 1200 wet and dry and household soap. When the vehicle is completely mat follow on with the 2A compound and cut back the surface.
22. Finally finish out by polishing with Tetrosyl T Cut which gives a very high gloss finish to the vehicle.
23. For extra deep lustre follow on with a good quality car wax. This waxing should be carried out at intervals of six months after the first application.

To keep the paint film in excellent condition the following method should be adopted:

1. Wash the vehicle thoroughly at least once a week. The washing process should consist of wetting the car with a low pressure hose, then washing using a bucket of hot water with a teaspoon of good liquid soap, as used for washing up. Wash the vehicle down from the top using a soft sponge.
2. Repeat the process, and ensure all the grit is washed away.
3. Rinse with cold water by hose at low pressure, or douse with a sponge.
4. Leather the car with a good quality chamois leather until dry.
5. Every six months, apply T Cut to remove the road film and surface that has been contaminated by fall out. Re-wax using a good quality wax polish.

This process will ensure that the paintwork on your vehicle will last and look extremely good for up to ten years without serious deterioration.

Hose the underside of the vehicle frequently as this will help dilute the salt and road grime that is thrown up. It is always better to keep the car as clean as possible so that rusting on the chassis can be kept to a minimum.

Painting a GRP Bodyshell

It may be that, having a vehicle that is out of the ordinary, the case for using metallic or even the new pearlescence colour may be made. If this decision is taken then only one consideration need be brought forward: the repair of these finishes is highly complex and the process requires both great care and thorough understanding. However, it is not an insurmountable problem and provided that this understanding exists then repairs can be satisfactorily carried out.

Metallic finish and basecoat and clear

Use all the basic preparation as previously explained, up to the application of colour. Then:

1. Follow the paint manufacturer's recommendations for thinning the colour. A cellulose metallic finish will thin out at between 23 seconds and 26 seconds BSB4 viscosity cup.
2. Apply the colour, starting on the roof and working down and round the vehicle. Ensure that all panels are assembled on the vehicle prior to spraying, as different effects can be achieved when panels are sprayed separately.
3. Apply at least three single coats with a gun overlap of up to 80 per cent.
4. For final finish apply a wet on wet coat. This means respraying the vehicle again panel by panel to increase the film build.
5. If a basecoat system is being used then apply the overlay lacquer in a time span of 5 to 15 minutes after the last wet coat of metallic finish.
6. Allow to through dry and then polish, as already stated.

Repair of metallic finish

It is advisable to retain some of the original finish in an *unthinned* condition in a place where the temperature will not drop below freezing. A cellulose colour metallic or straight will remain in usable condition for up to five years. It is important not to add thinners to any material chosen for retention, as it will cause the pigment to drop out of suspension which will make it almost impossible to use later.

After the damaged area on the vehicle has been correctly repaired then the process of repainting can begin. It is important to understand with metallics that for an invisible repair it is necessary to ensure that:

1. The colour matches.
2. The appearance of the metallic when sprayed matches.
3. The texture of the metallic matches the original.

The first item is out of your control unless you have a retained sample of the original or can purchase and use the correct tinters to obtain the colour match you require. The other two items are in your control as the one who applies the colour. Gun distance and speed of pass dictate the random orientation of the metal flake in the metallic and therefore affect the appearance and the texture. This is a skill that has to be practised and developed, and needs to be carried out first on a test panel. The use of blending clear helps the situation dramatically, when used correctly.

Vehicle Fabrications in GRP

Presuppose that a door has been damaged and the bodywork repaired:

1. Abrade the repaired surface with 320 wet and dry.
2. Wipe down with spirit wipe.
3. Apply a single coat of self-etch primer ensuring that it is not sprayed over original colour.
4. Apply three or four coats of cellulose primer filler, such as ICI RPF 800.
5. When through dry, spray a light black guide coat.
6 .Wet flat using a block and 600 wet and dry paper.
7. Dry off thoroughly.
8. Spirit wipe.
9. Apply three coats of correctly thinned colour over the damaged area only. Overlap each coat with the successive coat.
10. Allow to dry through.
11. Wet flat the whole door lightly with 800 wet and dry.
12. Apply two full coats of colour over the repaired area.
13. Add 25 per cent of blending clear lacquer to the contents of the pot on the gun.
14. Spray the primary area (Fig. 50).

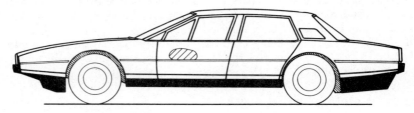

Fig. 50 Primary paint area.

15. Add 50 per cent lacquer to the contents of the pot.
16. Spray the secondary area (Fig. 51).

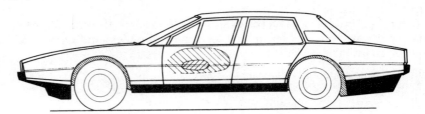

Fig. 51 Secondary paint area.

17. Add 75 per cent lacquer to the pot and spray the whole door.
18. Under no circumstances spray lacquer on its own as it relies on the colour to help stabilise on long-term weatherability.
19. Allow the paintwork to dry through and then polish as previously described.

Painting a GRP Bodyshell

In effect, by applying blending lacquer over the original finish in varying degrees of opacity, the original finish is allowed to show through, and therefore distracts the eye away from the repaired area.

12

Material Developments

The material developments of all glass reinforced polyesters appear to be on a preferred route to 'sandwiching'. Both the giant conglomerates, ICI and DuPont, appear to be making strides in this direction.

ICI Plastics Division have recently announced a new sandwich material said to be cheaper and lighter than sheet steel, for producing automotive panels, which has been developed in conjunction with Austin Rover in Britain. Two different materials are injected simultaneously into the mould. The substrate is a glass reinforced propathene modified polypropylene to provide strength and rigidity. The reinforcement enables the propathene to withstand the high stoving temperatures necessary for on-line painting. The other component is a thin layer of non-reinforced polypropylene to act as an outer skin. This obviates the usual problem with glass reinforced metals of an orange peel finish when painting. A Class A finish is possible and the surface will withstand temperatures up to 130°C.

The development has already resulted in an injected moulded car wing said to be significantly less expensive and lighter than those made from sheet steel. Developers say the method is economically suitable for manufacturing 200,000 components per year from one set of tooling.

For GRP and the thermoplastic materials to be used extensively within the original motor manufacturer's system they have to fit in with the on-line systems employed. Much investment has been made over the years in steel construction of motor vehicles and parts, and everything from fixings to paint finish systems revolves round the ability of sheet steel to perform in a predictable manner. For example, a steel body unit will happily take the panel temperature of 140°C to 160°C demanded by the paint manufacturers to crosslink thermosetting and thermoplastic acrylic finishes. At these temperatures GRP and composites are of uncertain performance and therefore either finish paint must be formulated to crosslink at a lower temperature, or the GRP and composites must be structured to cope with the higher temperature.

Volume of parts is another area that causes the concern. A motor manufacturer is always talking in terms of 200,000+ parts per annum and realistically the methods of manufacture employed in GRP and composites at present do not give these demand figures at the low unit cost. However, a great deal of research is being carried out into this particular facet and the GKN Sankey programme for road springs shows that this problem should soon be overcome. They plan to produce 600,000 units per annum by 1987.

Low volume manufacturers are in a very strong position because their annual production demands are low and the unit price has a broader margin. Aston Martin Lagonda Ltd produce four vehicles per week, and have moved many items from steel or aluminium into GRP. Now their own in-house production of GRP allows total flexibility and a very tight control on quality. Development of new models is aided by the GRP facility to form up shapes and

Material Developments

ducting on spoilers and air dams quickly for evaluation before a commitment to production is made. The obvious advantage of low weight plus strength is paramount to safety and shattering performance. The new Aston Martin Zagato has GRP fabrications fitted throughout the vehicle, and this has contributed to the weight-saving for performance, proved by the fact that the top speed of this most luxurious sports car is in excess of 180 mph.

Comparatively low cost tooling for GRP use is a prime factor in its adoption by the low volume manufacturers. Rolls-Royce, currently producing sixty vehicles a week, are using GRP fabrications within the vehicles. So many items can be fabricated and then, after fitting, forgotten about for the life of the vehicle. Internal pieces will never rot, rust or break down. They are complete and inert.

The lead taken by aviation and aerospace ensures a follow-on in the automotive industry. For example, the usage of composites in civil aircraft construction is expected to rise to 65 per cent by weight. These are staggering amounts and the investment put into the development will be extensive. The knowledge gained must serve the automotive industry.

A material development that is causing interest at present and which will show an ever-increasing spectrum of applications is that of glass reinforced polyurethanes. A major design attraction of reaction injection moulding (RIM) polyurethanes is the facility to tailor mechanical properties to suit an application. Polyurethanes are a family of materials and therefore do not present a single set of properties. This, coupled with high impact resistance, the ability to form complex shaped components with good surface finish that will take paint, and lower tooling and press costs, have seen RIM hailed by some as the best thing to have happened in the polymer industry in recent years.

Despite the wide range of RIM polyurethanes, they are characterised by relatively low moduli of flexural stiffness, typically from 150 MN/m^2 to 1,000 MN/m^2, although most interest is in the range up to about 850 MN/m^2. The introduction of glass reinforcing brings as much as a twofold increase on this, giving a real boost to the design appeal of reinforced reaction injection moulding (RRIM) polyurethanes. The glass also reduces thermal expansion coefficients to a level compatible with conventional engineering materials, and greatly reduces the heat sag. Such benefits, however, have to be traded off against some reductions in impact strength.

Although still relatively a new plastics technology innovation, RRIM is established and well-proven in commercial production. Its transition from the laboratory to the commercial moulder has drawn heavily on materials developed by ICI and Fibreglass UK Ltd and the commitment of a mere handful of UK moulders to establishing parameters of mould design and moulding techniques.

A further and quite decisive factor in establishing the commercial viability of RIM and RRIM has been the attitude of the motor industry which currently is the major outlet for the mouldings. At first, the industry took to the material for 'add-on' items such as spoilers, grills, wheel arches and the like, but interest is increasing in the material for body panels and more major components. Though the car of the future with a plastic clad light alloy space frame still remains some way off, RRIM looks like being a front-runner when it does

Vehicle Fabrications in GRP

arrive. High impact strength, lightness, acceptable stiffness, good surface finish and unexpectedly lower tooling costs, are what caught the eye of the automotive designer – all benefits that must surely recommend its use to other industrial sectors.

Development pedigree

RRIM polyurethanes do not fit neatly into the conventional thermosetting thermoplastics categorisation. They are not heat-reprocessable; their processing technique has far more in common with injection than compression moulding. Their properties lie between the typically representative values for the two groups. The two components of the polyurethane system, a liquid polyol and an isocyanate hardener, come together at high pressure in a mixing head which injects the mix directly into the mould. The high pressure mixing, typically at around 20 MN/m^2, ensures extremely thorough mixing through molecular impingement leading to much faster polymerisation than expected of conventional low pressure mixed polyurethane reactions. Demoulding cycles are similar to those of injection moulding but a flash trimming operation is inherent in the RRIM or RIM process.

Although the mixing is at high pressure, injection into the mould is at around 0.6 MN/m^2 so mould clamping forces are an order of magnitude lower than compression moulding and a factor of 50 down on injection moulding, of considerable import to press capacity and hence processing cost. As already outlined, the development of RRIM represents far more than a minor extension to the RIM process and the form of the reinforcement has been the subject of considerable interest and development investment in the UK by ICI and Fibreglass Ltd.

The original, and still widely used, reinforcement was hammer milled glass, in which fibre lengths range from a few millimetres down to dust. Hammer milled glass is far from ideal. Optimum fibre length is around 1.5 mm: any less than this gives poor matrix adhesion and allows fibres to pull out; longer fibres are prone to individual failure. With only a small proportion of the hammer milled fibre around optimum length, more bulk is needed to produce the same reinforcing effects as 1.5 mm fibres. This additional bulk is undesirable. First, its inconvenience in processing appears in pumping and mixing, as the filler is added to the polyol to form a slurry. Second, the greater the proportion of filler the greater is the reduction in impact resistance, which makes the material less resilient, and third, the levels of hammer milled glass needed to achieve good flexural moduli begin to detract from surface finish.

The ICI and Fibreglass Ltd solution has been a move to using chopped strands of close to the optimum 1.5 mm length. Perfecting this has involved ensuring that the strands do not clump together, do not suffer damage in the nozzle, and do not cause undue abrasion in the mixing head.

The achievement is that, for roughly equivalent reinforcing effect, 8 per cent chopped glass corresponds to 25 per cent hammer milled. At present these represent the maximum practical levels of filler addition, but future developments in materials and processing machinery hold out the prospects of higher levels of reinforcement in the foreseeable future.

Material Developments

Alongside all the advantages of RRIM in overcoming the typical limitations of RIM (low stiffness, a high thermal expansion, and poor heat sag performance), it must be stressed that the development of RRIM is complementary to RIM and RRIM will not displace RIM. Rather, the two together offer designers a material with an unrivalled range of properties which can be tailored to suit a wide range of applications.

Tooling costs

Tooling costs for RRIM components are substantially less than for injection mouldings. Even if made in steel, RRIM tools have less complex feed arrangements and their cost is around 60 per cent of that of an injection mould. Press costs are also reduced because of the lower internal mould pressures which translate into lower clamping forces. But tools, even production tools, need not be of steel. Cast aluminium and kirksite (zinc alloy) construction are perfectly adequate and can reduce tool cost to less than half that of injection moulding.

For small runs and prototypes, plain and spray coated epoxy tools costing from a fifth to a third of that of a steel tool (one eighth to one quarter that of the injection mould tooling) offer runs of a few hundred to a few thousand. Low volume therefore favours RIM and RRIM in terms of unit cost. High volume throws the balance back towards injection moulding. Critical volume depends on factors such as moulding size, but typically 1,000 to 20,000 a year strongly indicates RRIM rather than injection moulding.

Whilst the mould making requirements are in some respects less demanding than for injection moulding, the ground rules are less well defined. While familiarity will ensure that most toolmakers will experience little difficulty in an injection moulding tool, the quite different demands made by RRIM can cause problems.

The importance of mould design is naturally crucial to the quality of RRIM and RIM parts. The key consideration is that material enters the mould in the form of a fast flowing liquid rather than the very high pressure jet of molten plastics of injection moulding. This holds implications for gate design and mould venting. The material must enter the mould cavity and displace the air as it fills the cavity in a controlled manner. Turbulence created by poor gate design is disastrous, and entrapped air will give rise to uncontrolled porosity in the finished moulding. To ensure laminar flow, gates are usually bar- or fan-shaped and on the mould parting line, gate flow rates are kept around 1 to 2 m/s.

Mould temperatures, too, need careful control as the polymerisation reaction is strongly exothermic.

In sheer tonnage the US automotive industry leads the way in RRIM and RIM applications but the great majority is in low stiffness RIM for impact absorbing soft front ends of cars. European, and particularly UK, practice appears more advanced, with RIM and RRIM increasingly being used for panel work as well as grilles, spoilers and wheel arch extensions. Lotus, with its history of materials innovation in the motor industry, is using RRIM on its latest Excel. The entire lower half of the nose and tail are moulded in high

Vehicle Fabrications in GRP

modulus glass-filled RRIM. Improved styling and aerodynamics, including attention to engine compartment airflow which can make a surprising drag contribution, resulted in an extremely complicated shape for the nose section, which had to house the lamp clusters and number plate. RRIM, with its ability to form deeply dished, complex, smooth surfaced and lightweight components, presented the ideal choice.

Volume considerations weigh heavily in current automotive examples of RRIM applications but this does not rule out its popularity in the high volume sector. Low tooling costs offer the promise of much lower costs of component re-design, a factor assuming greater importance in the cutthroat competitiveness in the motor industry.

For example, Ford took advantage of RRIM to re-model the grille for its A Series light commercial, derived from the Transit, for continental markets. The Transit grille is injection moulded in ABS, commercial good sense with volumes in the 250 per week bracket, but a 3,000 per year A Series volume was much better suited to RRIM.

BL's successful limited production MG Metro Turbo and Rover Vitesse provide parallel examples. The wider wheels of the MG call for fixed arches; while the cost of re-tooling for special steel panels was prohibitive, RRIM eyebrow mouldings were a low-cost solution. The vehicle's spoiler is also moulded in RRIM.

Awkward shaped components, too, can be suitable for a change to RRIM, again especially where volumes rule out injection moulding. The rear light cluster for BL's Range Rover is made of a deep drawn steel pressing reinforced with two spot welded steel strips. A one-piece RRIM moulding is now being looked at as a replacement.

Another area of RRIM application is in prototyping. A low-cost epoxy tool can produce a few hundred units of a new design for proving. Minor modifications can be made before going on to the production version, which could be in RRIM but might, depending on volume, be an injection moulding.

Though the automotive field has to date been the only industry really switched on to RIM and RRIM, the attributes that endear the material here are equally attractive outside this industry, perhaps more so, as these volumes tend to be smaller. But it is in the motor industry that the greatest progress will be made in the future.

Multi-discipline plastics moulding

As OEMs worldwide concentrate on fewer suppliers, those with a range of plastic moulding capabilities, and with technical support in design and production development, are the ones most likely to survive. This is the view of the Skaraplast Division of the Swedish Perstorp group who claim to be in the forefront of a number of processing technologies and already direct 55 per cent of output to the automotive sector. Machines up to 2,200 tonne capacity allow large size mouldings, including radiator grilles and sun visors (one product is 2.3 m long) for heavy trucks.

RIM and RRIM products can be produced, as can thermoplastic mouldings. A joint development venture with Volvo has led to a new method of vac

Material Developments

forming PVC foils onto injection moulded components. The Volvo 760 dashboard offers the following formed in this process: electronic turn indicator, bulb failure monitor, lighting control arm, central processor, air-conditioning control module, light current switches, trip computer keyboard, lighting and clock module and the gearbox position switch.

A ten-year experience in RIM and RRIM has led to an ability to tailor products over a range of requirements between low and high density (40–1000 kg/m), flexible to rigid, with good high temperature performance (up to 130°C). An intermediate grille panel for the Leyland T45 range is a typical product. The cashiering process for laminating vac formed PVC onto ABS carrier mouldings has been adopted by the Volvo 760 car for body pillar mouldings and outlets are now being sought for fascias and door panels. Parts up to 800 × 1600 mm can be handled on presses producing up to 90 units per hour which recently have been developed to handle fabric coatings.

It is interesting to see in Table 4 how a supplier in both disciplines sees the relative advantages of thermoplastic and polyurethanes.

Table 4

Thermoplastics	Polyurethanes
Wide range, from touch	Very good thermo-insulator
Shock resistant to rigid	Low density
Surface hard	Good rigidity in relation to density to
Good damping characteristics for both	partweight
vibration and sound	Good resistance to abrasives
Low density	Varying wall thickness
Good resistance to corrosion	Low investment costs for large
Good resistance to chemicals	details
Very good electrical insulator	Very good outside properties
Very good thermo-insulator	
Glass clear	
Colouring possibilities unlimited	
Easy to mould	

DuPont recently introduced a new stiffened, super tough grade of Rynite thermoplastic polyester resin which is expected to find major applications in the automotive field. Designated SST, the grade contains a proprietary toughening system for polyethylene terephthalate polyester (PETP). A flexural modulus of almost 7 GN/m^2, and Izod impact value of 235 J/m, are exceptional for reinforced thermoplastics. The company claims that even at –40°C it is 50 per cent tougher than glass reinforced polycarbonate is at room temperature. Applications are seen in small body parts painted on or off-line, including grilles, fuel filler flaps and small panels, body panel reinforcements, structural body components, and sun roof frames. Also the new super tough thermoplastic can be used for various mechanical parts such as steering wheel

Vehicle Fabrications in GRP

reinforcements, steering wheel column supporters and rack and pinion housings.

Use of recently introduced Dymetrol fabric enables the creation of lightweight thin profile seat systems, employing new construction methods. These fabrics are woven from high strength elastomeric monofilaments in one direction, and high performance textile yarn in the other. The former damp out road vibrations and provide resilient passenger support, while the latter maintains fabric durability. The fabrics are bonded at the monofilament yarn intersections to provide high tear strength and dimensional stability, resulting in a breathable, odourless fabric of considerable strength.

Without doubt, DuPont have made most significant developments in all aspects of the motor industry plastics. The following are some examples of the most modern materials and their automotive uses.

Delrin

This is the world's first acetal resin. It is a highly versatile engineering plastic with metal-like properties. It offers outstanding strength, stiffness and hardness, low wear and low friction, dimensional stability, fatigue abrasion and solvent resistance, and is self lubricating. Now DuPont have achieved a major breakthrough in polymer chemistry with a new generation of super tough and toughened acetal resins: Delrin ST and Delrin T. Delrin 100ST is seven times tougher in notched Izod and falling weight testing than standard Delrin at 23°C. In impact fatigue tests it is two and a half times more resistant than chemically toughened polyester and up to ten times more than present polycarbonates. Volkswagen use Delrin 100 for a fuel pump housing. It works in methanol and other alcohol fuel blends. The six components of the pump use snap-on fasteners and self tapping screws for easy and economical assembly.

Zytel

Zytel nylon was first introduced over thirty years ago and has become the largest selling engineering plastic. Flame retardant Zytel FR 10 offers flexibility and an exceptional degree of elongation. This makes it ideal for parts with thin sections such as electrical connectors. AMP use Zytel for automotive electrical connectors. Insulation properties, self-extinguishing flammability rating, and excellent flow properties in multi-cavity moulds, plus toughness, are important for the moulded hinge. Zytel ST has outstanding resistance to chemicals, solvents, oils and greases. It is more useful in problem environments than polycarbonate. It can survive temperatures of up to 170°C in paint ovens. Jaguar Cars use Zytel ST to produce the cooling fan. It cut original steel weight by one-third, and eliminated costly spot welding, painting and balancing. It is tested to 9000 rpm in temperatures from –40°C to 125°C.

Glass reinforced Zytel is used for items needing to cope with high deflection temperatures. An interesting development of Zytel 70G-30 HSL-R (hydrolysis resistant and heat stabilised glass reinforced Zytel) is the windscreen washer water heater called 'hot jet'. Produced by Rold Elettromeccanica, Italy, this

Material Developments

component heats the water to 60°C within a few seconds – the required temperature for perfect windscreen cleaning.

The ability to withstand fatigue resistance in the presence of anti-freeze at 120°C under pressure of up to 2 bars made GRZ the ideal choice for radiator end caps. European car manufacturers have appreciated the weight-saving and the lower price per unit.

Renault, Volvo and Citroën use GRZ for rearview mirror housings. It is attractive, tough and has high creep resistance round and in socket joint. High heat deflection is important in the re-touching ovens.

Minlon

Minlon is a nylon that is reinforced with mineral or mineral glass particles. It displays excellent strength and stiffness allied to outstanding heat resistance. It will withstand paint oven temperatures and is claimed to be the ultimate resin for processability and low warpage. Flame retardant Minlon FR 60 also contains these advantages. In addition, its stability has been demonstrated at processing temperatures of over 300°C. The on-line paintability is a must for many plastic body applications. Furthermore, sometimes particular impact resistance characteristics are required. Minlon 13 T1 was chosen by Peugeot for the front and rear spoilers for the '205' model. When AMP redesigned the Volkswagen electrical junction box they used Minlon and saved 45 per cent weight. Freedom from warp and dimensional stability were of paramount importance for the printed circuit board technology.

The air intake grille on the Fiat Ritmo was the first plastic part to be painted on-line in Europe. It shows the ability of Minlon to replace sheet metal or die-cast aluminium, and shows how to reduce weight without loss of performance.

Bexloy

Bexloy is a new family of engineering plastics designed for the moulding of large vertical exterior car body panels and bumpers. Bexloy 'C' is the first resin of this new family of products. It is the result of several years of development and marks a significant addition to DuPont's nylon technology. It is based on a new amorphous polyamide invented by the company's research and development department.

The first commercial application for Bexloy was the blow-moulded rear spoiler of the Pontiac Indy Fiero in 1984. DuPont's automotive resin was selected as it weighs only half the SMC alternative. This allows lighter and less costly support of the spoiler. With this innovative design, tooling and assembly costs were significantly reduced.

A bumper unit under test has a Class A surface out of mould and withstands 8 km/h impacts at –30°C. Bexloy resins out-perform RIM and SMC in standard ASTM Izod impact tests. They give high energy absorbing ductile breaks at room temperature, with values for some resins exceeding 1065 J/m. Several formulations tailored for bumper applications give ductile breaks at temperatures as low as –30°C.

Bexloy has both better stiffness and higher impact strength than RIM and

77

Vehicle Fabrications in GRP

RRIM. It can therefore be formulated to provide a useful combination of properties for a variety of automotive applications. For most Bexloy resins stiffness remains relatively constant over the range –30°C to 75°C. In bumper panels of 3 mm thickness, it provides the needed stiffness to replace steel in bumpers and reduces weight by almost one half.

Parts made from Bexloy retain excellent stiffness at paint oven temperatures. Some have deflection temperatures as high as 158°C at 0.5 MPa, and several retain a flexural modulus of 840 MPa at 140°C. This means that Bexloy can be top coated on-line, side by side with steel panels, while using conventional equipment and paint systems. Bexloy panels resist stress cracking from common automotive fluids and road chemicals under a wide range of environmental conditions. Tests have shown that it has excellent resistance to hydrocarbons, unleaded fuel, transmission fluids and toluene. Special formulations resist alcohols and glycols, even in high stress situations.

The acceleration of the technology now available will surely bring the 'all-plastic' vehicle to the public within the near future. The development from the early days of glass fibre up to the present gives only a glimpse of what is possible. The exciting future of composites and all the derivatives ensures a better and safer motoring environment for everyone who is connected in any way with transportation.

Surely, these remarkable materials will make their presence felt in the Third World. Trucks and light commercial vehicles that will not rust and are virtually indestructible will be of the greatest asset to these emerging nations. As science and technology press forward, the whole world population must benefit in the long run.

13
Future Trends

The development of all types of glass reinforced polyesters looks certain to continue at an accelerating pace towards the next century. Without doubt, the aerospace industry has had, and will continue to have, an enormous influence on the development of these materials.

The only problem facing the car manufacturers at present is the quantity of the items required for mass production. However, it can only be a matter of time before this problem is finally overcome. As the demand for longer and longer warranty periods forces the competition within the market place, so it is certain that a great deal of GRP will be used in the main construction of the popular motor vehicle.

At this point in time it is hard to imagine a total GRP vehicle from the major manufacturers but it is likely that the development of more sophisticated resins will enable the final transition to the 'all-plastic' car.

The development of further vehicle parts in GRP is already well under way and the trend is likely to continue. GKN Composites Ltd have announced that they are starting the manufacture of glass fibre reinforced epoxy resin road springs for light commercial vehicles. A totally new plant has been laid down at Telford in Shropshire to construct these parts, and by 1987 the output is expected to reach 600,000 springs per year. The company is carrying out an intensive research project into passenger vehicle springs and it is certain to develop into other areas.

Engineering use of thermoplastics is now increasing annually by approximately 8 per cent. Batch moulding techniques and alloying processes promise more substitutions of plastics for steels.

Some designers feel that steels and light alloy castings will face very tough competition from thermoplastics in the future. According to a British Plastics Processing Working Party Report, *British Plastics, The Next Ten Years*, the most obvious growth of plastics will occur in the construction of automobiles. In 1980 the average European passenger car used 5kg of thermoplastics. By 1990, it will be over 98 kg per car. From the automobile's 1980 consumption of around 20 per cent of the world's output of plastics, this figure is anticipated to be over 50 per cent by the year 2000.

Not only will the growth take place in bumper, spoilers and body parts, but research at present shows growth in areas such as carburettor housings, fuel pumps, rocker covers and in the main and sub-chassis. Gearboxes and clutch housings may also turn to thermoplastic construction.

The considerable experience that Lotus has gained in composites through its own body manufacture led to the development of the vacuum assisted resin injection (VARI) process, first used in the 1970s on the lotus Elite. This process is now under further development to produce a full showroom finish including a paint system straight from the mould. This process, known as VARIP (the P standing for paint), is to be used on the new X100 sports car and it will remove

79

Vehicle Fabrications in GRP

the need for final painting. This is an interesting new development and it will be interesting to see the repairability process that will have to be applied to refinish the vehicle to give the same cosmetic appearance as the original finish.

Prior to the composite lay-up the mould is polished with a material developed for Lotus by ICI. Paint is then sprayed onto the mould surface before the dry fibre matting is laid into the mould. The male, or top of the mould, is then clamped in place and the gap between the two halves evacuated. Resin is allowed to be drawn into the mould by the vacuum impregnating the fibre. After a curing period the mould is opened and the product is removed, complete with its finished paint surface bonded to the panel.

The new Lotus Etna is a fully stressed FRP unitary structure produced by the VARI process. It comprises a top and bottom moulding which has an integral foam-filled beam section to ensure ample stiffness and impact resistance. Steel subframes carry the drive line elements and the suspension and steering systems, and the superstructure is a carbon fibre Kevlar composite cage clad with the tinted glass. Polyurethane RRIM is used for the front and rear sections which incorporate the bumpers. This mid-engined Lotus is expected to be in production during 1988.

Another manufacturer that has in the past contributed to the development of GRP bodied vehicles is the Reliant Motor Company. After more than twenty years of GRP bodybuilding, Reliant have plenty of in-depth experience in this area. However, the production rate of their new SS1 sports car was intended to be higher than that of the previous, larger and more costly Scimitar range, so Reliant looked for more modern techniques than the well-tried and traditional hand lay-up manufacturing process. After a most thorough evaluation is was decided to use no fewer than four different methods, each the optimum for the particular body component.

Michelotti's styling brief covered the use of separately moulded body panels with overlaps to conceal the attachment points. This scheme gave three significant advantages:

1. Panels could be bolted directly to the steel tube armatures extending from the main frame work;
2. Panel manufacture was simplified by the ability to reduce moulding size;
3. Damage repair in service would be cheaper, quicker and easier than with a small number of large mouldings.

First of the four building methods is the original polyester glass hand lay-up; used for the boot cavity moulding inner body, headlamp surround and doors. The inner body is the largest single moulding, comprising the floor, tunnel cover, front and rear bulkheads and steel tube reinforced windscreen frame.

The second method is the RRIM (reinforced reaction injection moulding) which produces the polyurethane semi-flexible front and rear bumpers, and the front and rear wings, giving excellent impact and damage resistance. These mouldings are supplied by Dunlop GRG's Engineered Plastics Division.

The new development VARI, mentioned previously, is used for the bonnet skin. This system sandwiches a core of rigid urethane foam, resulting in a light, stiff and noise absorbing panel, the functionality of which is enhanced by another novelty, an intumescent fibre barrier coating on the inner surface. The

Future Trends

process for producing the bonnet is a joint development of Reliant and Scott Bader, their long-time collaborators on the plastics side.

Lastly, the boot lid is a cold pressing in polyester, with both mineral and glass reinforcement to achieve very great stiffness. Most of the body panels are fixed by Taptite bolts, giving positive location with quick assembly. The windscreen is directly glazed and the front quarter lights are fixed.

The use of these developed materials clearly highlights current trends – the mixing of a range of polyester materials best suited to the design part of the vehicle. This is how the use of plastics will invade the traditional steel monocoque body shell – slowly and surely, as each material is identified as highly worthwhile for a whole range of technical reasons.

In the USA, development of the all-plastics vehicle is moving forward at an ever-increasing pace and the design and construction of the total plastic car is coming nearer. Next to Chevrolet's Corvette, the 5 litre V8 GM engined Avanti is the oldest continuous production composite bodied vehicle. The performance of that material can be credited with aiding the car's success over the last 23 years. Recently the design has been given its first facelift, but the styling department's issued dictum that weight is the enemy received even greater attention. Riding an overall 5 cm lower to the ground, the car has all its external body panels changed from contact to compression moulding, employing a low profile resin 'wet mix'.

It is in this area that additional and far-reaching developments can be reported for all applications of this type. Although based on automotive SMC technology, the claimed and demonstrated advantages of this new material are: much more consistent out-of-mould Class A surfaces, and more reliable physical properties. Ripples and long-term waviness, another problem with large, smooth-surfaced SMC parts, have also been successfully overcome. Those gains result from a significant difference in the glass reinforcement which, during moulding, does not flow in a manner that changes the chosen orientation.

Now supported on a full ladder steel cross member frame, the Avanti underbody is one of the largest matched metal die FRP mouldings yet produced. Glass fibre preforms, contoured to the mould surface for precise reinforcement placement, receive additional glass mat overlays to provide enhanced strength in critical stress areas. By ensuring throughout all components an average 30% minimum glass, all the strength to weight requirements are satisfied, without the use of other than a matrix of mineral-filled, general purpose polyester resin.

After moulding, semi automatic equipment and a specially formulated robot-dispensed urethane adhesive bonds the body shell components together prior to application of a urethane-based finish coat to the desired colour. Integral Kevlar reinforced safety bumpers, and cold press moulded bucket seat shells complete the reinforced plastics employed to achieve an all-up weight of 1700 kg. Equipment and internal fittings fully complement the specification of this luxury high performance hand-built sports car.

The panels for the upper parts of the Pontiac Fiero bodywork are, like the Avanti, compression moulding, but in an SMC formulation incorporating thermoplastic and rubber additives. This low profile material, plus an in-mould

Vehicle Fabrications in GRP

applied coating, is Pontiac's alternative for achieving the Class A surface. However, the lower panels are reaction injection mouldings of reinforced urethane, providing recovery properties in areas subject to abuse. Several employ 10 per cent glass fibre reinforcement, while others 22 per cent glass flake, to give improved surface quality and uniform mechanical properties in all directions. The largest SMC part, at just under 12 kg, is the upper quarter panel having overall dimensions of 173 × 135 × 152 cm. The largest glass flake RRIM part at just over 3 kg is the nearly 4 mm thick door outer panel, 122 cm wide by 50 cm high.

Though small under-bonnet reinforced plastic mouldings are now well accepted, major bodywork advantages have formerly been restricted to the specialist vehicle manufacturer. Both Avanti and Pontiac, and not forgetting one European example, the Citroën BX, move the possibility of the all bodied composite mass-produced car a significant step nearer.

It is likely that the major motor manufacturers will announce a small designated model to go all-composite. One can imagine Ford developing a small front-engined model in the Fiesta mode as a market test-piece to experience the reaction of the public at large to this development. It could be a simple and attractive design using the mechanical components of an already-existing model. It is likely to be a great success, as a very attractive body warranty could be offered. An indestructible car as far as use and elements are concerned would be a great marketing feature. However, the fact that mechanical spares would have to remain on offer indefinitely might temper the extreme long-term position.

The trend exists and will continue to grow until the whole of the mass-produced market is offering a total, or at least part composite construction. It could be, for instance, that manufacturers such as Volvo would be reluctant to release themselves from the extremely strong steel cage which is the backbone of their vehicles and which offers the full safety element in the event of collision. However, it is probable that when Kevlar and future developments of this type can be offered to the motor manufacturers in a form that can be incorporated into mass-production, Volvo and other safety-conscious manufacturers might then proceed down that route. It is vitally important that great emphasis is placed on passenger safety, and Volvo, along with Mercedes, must be applauded for their concern in this area. Both manufacturers have made a great sales aid out of safety features, and rightly so. New materials will enhance this safety programme.

14

Repair of a GRP Bodyshell

There are basically three types of damage that need identification for repair purposes.

1. A small gash in the gel coat
2. A damaged area that has caused a hole in the laminate.
3. A damaged area so great that a new moulded piece needs to be fitted to the body, for example, a new front wing.

Small gash

Check round the area for signs of stress and if present then relieve them as previously described by drilling the extremities. Then use the catalyst body filler to fill the gash and the stress holes. Rub down by hand or block with 80 grit production paper and finish with 320 wet and dry paper. This surface will be good enough to paint on.

Hole damage

Examine the damage carefully from both sides, and if it is in an area where road dirt has been thrown up then clean the underside very thoroughly. Cut away the damaged laminate and get back to firm and undamaged laminate. Dry rub the underside area to give adhesion and wipe with acetate or cellulose thinners to open the surface a little to receive fresh resin.

Over the outer side of the panel lay a thin plastic or form a mild steel sheet to the contour of the area to be repaired. Then from underneath coat up with gel coat and follow on with laminations in the normal way. Use a stipple brush to force the resin through the mat and follow on with the roller. When this has cured out then remove the steel or plastic former and abrade the surface with production paper. Any filling that is necessary should be carried out using the catalyst filler.

Large damage area

If a large area has been badly damaged then the total removal of that area is the correct procedure. If this panel was the front wing of a vehicle, for example, then the whole panel should be cut away at a point beyond the stress area and convenient for re-laminating (Fig. 52).

When the damaged panel is removed, ensure that the reverse side is cleaned of any road dirt, grease or other contaminate. Abrade the surface with 40 grit production paper and clean down with acetate or cellulose thinners.

A new moulded panel must be acquired and trimmed up ready for fitting. It is most desirable to get an edge to edge fitting on this panel (Fig. 53). Once this

Vehicle Fabrications in GRP

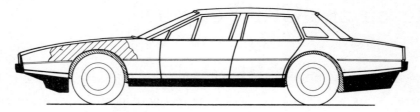

Fig. 52 Extensive wing damage area.

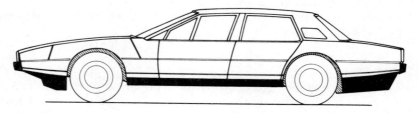

Fig. 53 Edge to edge wing repair.

has been achieved then the panel must be held in place by fixing with battens which can be clamped or screwed into position.

Next, apply gel coat and work it into the joints, followed by three laminations applied in the normal way. To stop the gel coat from passing out onto the surface of the panels apply masking tape down the seams. When the whole piece is fully cured, remove the tape and battens. Then dress the surface with 80 grit production paper. The panel can then be re-painted. It is possible to repair any GRP vehicle in this way, and it is also possible to return a bodyshell to the original mould and lay up a large area without difficulty.

When returning a complete shell to a mould, it is important to bear in mind that this is a last resort situation, because of the damage that may occur. Ensure that the damaged shell is as clean as possible and that all the preparation work has been fully carried out. On no account attempt repairs on the original shell once it is in the mould. The shell must be gently lowered in and a very tight close fit must be obtained before applying gel coat. There is a tendency for the gel coat and subsequent resin applications to run under the bodyshell that has been placed in the mould.

Once a close set into the mould has been achieved then lay in gel coat and follow up with the normal three laminations of $1\frac{1}{2}$ oz mat. Ensure that a good overlap of material gives the effective bond to the 'old' shell. Allow the normal cure out time and then release from the mould as usual.

Allow the body to stand before attempting the work necessary to make an invisible join, i.e. flatting back the join area with production paper. Repainting may take place after the 'green period' has passed and the shell is fully cured out.

Sometimes it can be more effective, and quicker, to make a complete shell

Repair of a GRP Bodyshell

rather than attempt a repair. It does depend on the type of damage sustained and the extent of that damage, as often stress has been raised and it will arrive at a later date within the shell.

15
Safety

The safe handling of all GRP materials is of prime importance and under no circumstances should any reductions be made in the safety procedures, for whatever reason.

Be properly equipped, store the materials safely and correctly – or do not attempt to work in this medium.

Very large quantities of organic peroxides are being used as initiators for the curing of unsaturated polyester resins. To a greater or lesser extent, organic peroxides are unstable compounds. Therefore it is necessary to be aware of their potential hazards and of the need to take greater care in handling and storing these substances. Nevertheless it should be emphasised that the commercially available organic peroxides are safe to use when handled properly.

Flammability

All organic peroxides should be considered as flammable substances and even in some cases the presence of air, i.e. oxygen, is not essential for combustion. These latter organic peroxides are explosively combustible. Once ignited, most organic peroxides burn fiercely.

A well known method by which to characterise the flammability of liquids is to determine their flash point. The obtained value refers to the lowest product temperature at which the vapour air mixture can be ignited. However, the flash point of an organic peroxide is only meaningful when it is below the temperature at which the organic peroxide starts to decompose essentially due to its thermal sensitivity. It should be noted that the flash point of an organic peroxide does not give any indication as to its hazardous nature or its stability.

Thermal sensitivity

From a safety point of view the thermal sensitivity of an organic peroxide is a very important parameter. Most of the commercially available organic peroxides used in the unsaturated polyester resin industry are reasonably stable at normal room temperature (approximately 20°). However, it is possible that at a slightly higher temperature the organic peroxide will already noticeably start to decompose. During this decomposition, which is an esothermic reaction, heat is released. If this decomposition proceeds so quickly that the decomposition heat is only partly dissipated, an accelerating increase in the product temperature will take place, finally resulting in a distinctly perceptible self-accelerating decomposition of the organic peroxide. Dependent on the circumstances, e.g. the quantity, the degree of confinement, etc., fierce decomposition, self ignition or even explosion may occur. It should be noted

Safety

that in the case of a flameless decomposition, the evolved vapours are flammable and hence any ignition source can ignite the vapour causing an explosion.

The temperature at which the decomposition may turn into a runaway reaction is different for each organic peroxide and depends also on the circumstances, e.g. the amount of peroxide, the ambient temperature, etc.

Contamination

The stability of organic peroxides is seriously reduced by a wide range of chemical compounds. Therefore contamination by, or contact with, heavy metals, iron, copper or their compounds, acids, alkalis and reducing agents, must be avoided. It should be remembered that the exothermic decomposition of a contaminated organic peroxide may take place at a considerably lower temperature than that of uncontaminated material. Care should be taken that under all circumstances direct contact with accelerators, which are normally used in combination with organic peroxides, does not occur, otherwise a violent decomposition or even an explosion may result.

Mechanical sensitivity

This includes the effects of impact, shock and friction. Most of the commercially available organic peroxides show a low degree of mechanical sensitivity, which means that no hazardous situations will arise provided that severe friction, causing an increase of the product temperature, rough handling and heavy impact, are avoided.

Ingestion and inhalation

In general organic peroxides are moderately toxic upon ingestion and inhalation. Owing to their corrosive character many organic peroxides can inflict severe injuries to the digestive organs on ingestion and to the respiratory tract on inhalation of the vapours.

Contact with the eyes

Although the effect varies according to the type, most organic peroxides are very dangerous in contact with the eyes. For instance, investigations have shown that ketone peroxides, e.g. Butanox M 50, and hydroperoxides, e.g. Trigonox AW 70, are particularly aggressive, causing extensive damage which may lead to blindness. On the other hand dibenzoyl peroxide, e.g. Lucidol KL 50, causes only temporary damage after short contact. In order to avoid any risk during the handling of organic peroxides, reliable eye protection must be worn. Use safety goggles at all times.

Contact with the skin

Owing to the corrosive nature of many peroxides, direct contact with the skin

Vehicle Fabrications in GRP

must be avoided. Skin contact may lead to lesions or irritation of an allergic nature. Always use protective gloves.

Take all measures necessary for bodily safety. Wear safety goggles, use protective gloves or a suitable barrier cream, do not inhale organic peroxide vapours, and always work in an adequately ventilated area.

Storage

The stability of organic peroxides is highly dependent upon temperature. Therefore the storage temperature of organic peroxides is very important for both safety and quality reasons. In the unsaturated polyester resin industry organic peroxides which should be stored at between 10°C and 25°C are normally used. For quality reasons a storage temperature below 25°C is advised, although with respect to safety, a storage temperature not exceeding 30°C, for several hours, is acceptable. In a climate where excessive temperatures may occur, special measures must be taken in order to avoid too high storage temperatures.

Glass fibre products

All fibres will irritate the skin. Wash with running water before applying soap and then wash very thoroughly. Always use a respirator when machining finished laminate. Always minimise the dust both in storage and work areas. Eliminate sources of ignition, including sparks from the discharge of static electricity. Take all the necessary precautions to protect the person and the environment from the dangers of glass fibre products. These modern materials are a most exciting development and they provide an excellent medium for the fabrication of a whole range of items; they do, however, demand careful and safe handling.

Glossary of Terms

Catalyst/hardener: An organic peroxide which must be added to liquid pre-accelerated polyester resin to make it set.

Catalyst filler: A body filler that hardens by mixing an organic peroxide with it. It is used extensively in the vehicle refinish industry.

Cure: Complete hardening of the moulding. This process starts very quickly, but can take several months for final hardening.

Female mould: An internal mould into which glass fibre mat and resin are laid up, producing a male moulding with an exterior finish.

Former/pattern: Terms applied to the original shape from which a mould is taken.

Gel coat: The resin which is applied to the mould first and thus becomes the exterior coat of the moulding.

GRP: The initials of glass reinforced polyester.

Laminate: (Noun) The layers of GRP. (Verb) To lay up GRP with resin.

Laying-up: The process of impregnating and consolidating the matrix of glass fibre mat and resin.

Male mould: An internal mould over which the GRP is laid up, producing a female moulding with an internal finish.

Monocoque: The name given to a complete vehicle shell including the floor structure.

Moulding: The finished article produced from the mould.

Release agent: Resins are excellent adhesives and will stick to most surfaces unless a release agent is applied.

Wetting out: Glass fibre reinforcements are held together with a chemical binder which dissolves in the resin. It is imperative that the reinforcement is thoroughly impregnated with resin and all of the binder dissolved to achieve full strength.

Index

accelerators, 4
Aston Martin Lagonda, 19, 39, 70, 71
average accident, 17

BIA (Thatcham), 17
Bexloy, 77
bolt tension, 59

carbon fibre, 2
catalysts, 4
chassis (steel), 7, 9
clay, 10, 11, 21
compressed air, 53

Delrin, 76
Dexion, 25, 26, 47
distortion, 34
DuPont, 70, 76

exothermic, 13, 32

fasteners, 55
female (mould), 11
fillers, 6
fire retardant resin, 12, 38
fish eyes, 64
force dry, 33

gel coat, 32
gel time, 5
glass fibre, 2
green time, 34

hardening time, 5

ICI, 70, 71, 80
inhibitors, 6

Kevlar, 48

laminations, 14, 28, 29, 30, 31, 50, 51
leaching, 64

male plug, 21
mat, 32, 38, 46
maturing time, 5
Minlon, 77
monomer, 3, 5
mould, 11, 12, 21, 22, 23, 28, 29, 30, 31, 33, 36

organic peroxides, 4

PVA (polyvinyl alcohol), 25, 26, 28, 30, 38
painting, 65
pigments, 6
plastic, 1
plug, 25
polyester, 3, 23
polyurethane foam, 12, 15, 75
pre-accelerated resins, 4

RIM, 71, 72, 73, 74
RRIM, 71, 72, 73, 74, 78
reinforcement, 56
reinforced moulds, 26, 47
release agents, 32
resin, 1, 2, 3, 4, 5
rovings, 13

SR surfacer, 28, 29
safety (materials), 86, 87, 88
split mould, 23
spotting, 64
stress relief, 62
styrene, 3

temperature, 8
thermoplastics, 75
thickness (GRP), 8

wax release, 30

Zytel, 76